THE ANATOMY OF KNOWLEDGE

THE ANATOMY
OF KNOWLEDGE

Papers Presented to the Study Group
on Foundations of Cultural Unity

Bowdoin College
1965 and 1966

Edited by
MARJORIE GRENE

UNIVERSITY OF MASSACHUSETTS PRESS

Theodore Lownik Library
St. Procopius College
Lisle, Illinois

First published 1969
by The University of Massachusetts Press
Amherst, Massachusetts
Printed in Great Britain

Copyright © 1969
by Marjorie Grene for the Study Group

121
.08
S933a

All rights reserved

No part of this book may be reproduced
in any form without permission from
the publisher, except for the quotation
of brief passages in criticism

Library of Congress Catalog Card Number 68-19672

CONTENTS

FOREWORD

The Study Group on Foundations of Cultural Unity was organized in 1964–65 by Michael Polanyi, Edward Pols and the present editor. With the help of grants by the Ford Foundation to Bowdoin College, Brunswick, Maine, which we here gratefully acknowledge, we held two week-long meetings, in August 1965 and 1966, both at Bowdoin. A statement of our general purpose follows this foreword, and the present volume collects a number of the papers which were discussed during the '65 and '66 conferences.

The Study Group has since changed its name to 'Study Group on the Unity of Knowledge'; with an enlarged committee and again with the generous support of the Ford Foundation, it is now based at the University of California, Davis.

Several of the papers included here have been published elsewhere. Professor Pantin's paper, which was written for our first meeting, has been included in his posthumous work, *The Relations between the Sciences*, Cambridge University Press, 1968. His untimely death was a heavy blow to our hopes for a rapprochement between biology and philosophy; we should record here both our sense of loss and our gratitude for his presence at the beginning of our enterprise. Professor Silber's paper, which was written for our 1966 meeting, appeared in a somewhat different form in the *Chicago Law Review* in the autumn of 1967. It was also reprinted, without due acknowledgment of our priority (this through no fault of the author) in *Phenomenology in America*, Chicago: Quadrangle, 1967. Professor Polanyi's 'Message of the Hungarian Revolution' appeared in *The American Scholar*, 35 (1966), pp. 661–76; their permission to include it here is gratefully acknowledged. His 'Structure of Consciousness' appeared in *Brain*, 88 (1965), pp. 799–810. The papers by Wigner, Pantin, Koch and Stallknecht will also appear in a *Psychological Issues* monograph containing a partial record of our first year's discussions; we are grateful to the editor, Dr. George Klein, for permission to publish them here, and also to the University of

vii

North Carolina Press for permission to include Miss Sewell's 'Cosmos and Kingdom,' which forms part of her second volume of collected poems, *Signs and Cities*, Chapel Hill: University of North Carolina Press, 1968.

Thanks are due to Mrs. Maria Jackson Parker, in Brunswick, Maine, for her work in duplicating the papers for our meetings, and to Mrs. Diana T. Hewitt in Davis, California, for her editorial work in preparing the manuscript for the press.

STUDY GROUP ON FOUNDATIONS
OF CULTURAL UNITY

Since the seventeenth century the kind of knowledge afforded by mathematical physics has come more and more to furnish mankind with an ideal for all knowledge. This ideal also carries with it a new conception of the nature of things: all things whatsoever are held to be intelligible ultimately in terms of the laws of inanimate nature. In the light of such a reductionist program, the finalistic nature of living beings, the sentience of animals and their intelligence, the responsible choices of man, his moral and aesthetic ideals, the fact of human greatness seem all of them anomalies that will be removed eventually by further progress. Their existence—even the existence of science itself —has no legitimate grounds; our deepest convictions lack all theoretical foundation.

This movement claims to unify science and to comprehend in it all subjects of study. But, since its ideal is fundamentally mistaken, the result has been to debase the conception of man entertained by the psychological and social sciences and at the same time to isolate from science the humanistic core of history and criticism. It has displaced the traditional endeavour of philosophy to comprehend the whole domain of human thought and produced instead distortion and fragmentation.

Although these views have been developing since the Copernican Revolution, they have gained the power to shake the foundations of our culture only in the last hundred years. There have been counter-movements: attempts to restore metaphysics, attempts to reformulate our conception of knowledge. In our own time, existentialism, supported by the ideas of phenomenology, has been perhaps the most potent counter-movement. Movements of this sort are strongholds defying the current scientific outlook; but they do not appear to be equipped for overthrowing and replacing it.

Today, however, there are signs of new centers of resistance among scholars, scientists, and writers in almost every region of knowledge. Many share the conviction that a deep-seated philosophical reform is needed—one that would radically alter prevailing conceptions, not only of the nature of knowledge and of creative achievements in general, but of the human agent who inquires and creates, and of the entire fabric of the culture formed by such activities. Of these thinkers, some have thus far confined their efforts to the critique of scientism; others are interested in the bearing of empirical investigations on the gradual clarification of epistemological problems; still others have embarked upon comprehensive philosophical reorientation.

Convinced that there is an unsuspected convergence of ideas separately developed in various fields, we propose a meeting of a number of persons who actively oppose in their work the scientism, and the related methodological and ontological over-simplifications, which in one or another form are ascendant in every field of scholarly and creative endeavor. The realization on the part of those attending this meeting that they are participating in one common movement of thought could strengthen each of them. It would help any attempts at systematizing the principles they perhaps unwittingly share, and would make it possible for the general ideas that emerge to receive criticism from workers in widely different fields.

It is proposed that the speakers at the meeting give an introduction to their personal views about the general questions before the group. For many this might best be done by reporting on their current work on some special subject that they are dealing with in the spirit suggested by the foregoing remarks. This does not mean that speakers should not also feel free to deal with the problems in more general terms; the point is merely that such a general approach is not necessary.

ORGANIZING COMMITTEE:
MICHAEL POLANYI (*Chairman*)
MARJORIE GRENE
EDWARD POLS

PARTICIPANTS AT THE MEETING

August 23-28, 1965:

Dr. M. R. A. Chance, Uffculme Clinic, Birmingham, England
Professor Frederick J. Crosson, Notre Dame University
Mr. Herbert Gold, San Francisco
Professor W. Haas, University of Manchester
Professor Marjorie Grene, University of California, Davis
Mr. Jerald C. Kindred, General Electric Company, Santa Barbara
Professor George S. Klein, New York University
Dr. Sigmund Koch, Ford Foundation
Mr. J. R. Lucas, Merton College, Oxford
Professor Henry A. Murray, Harvard University
Professor C. F. A. Pantin, Cambridge University
Dr. M. H. Pirenne, Oxford University
Professor Helmuth Plessner, Zürich
Professor Michael Polanyi, Oxford
Professor Edward Pols, Bowdoin College
Professor William H. Poteat, Duke University
Professor William T. Scott, University of Nevada
Miss Elizabeth Sewell, Salisbury, England
Professor John R. Silber, University of Texas
Professor Newton P. Stallknecht, Indiana University
Dr. Erwin Straus, Lexington, Kentucky
Professor Richard G. Tansey, San Jose State College
Professor Charles Taylor, McGill University
Professor Donald Weismann, University of Texas
Professor Eugene P. Wigner, Princeton University

August 21-27, 1966:

Mr. J. O. Bayley, New College, Oxford
Dr. Gotthard C. Booth, New York
Professor William H. Bossart, University of California, Davis
Dr. M. R. A. Chance, Uffculme Clinic, Birmingham, England
Professor Barry Commoner, Washington University, St. Louis

Professor Lon Fuller, Harvard Law School
Professor Marjorie Grene, University of California, Davis
Professor Hans Jonas, New School for Social Research
Professor Jerome Hall, Indiana University Law School
Dr. Torger Holtsmark, University of Oslo
Rev. Stanley L. Jaki, Seton Hall University
Mr. Jerald C. Kindred, General Electric, Santa Barbara
Dr. Sigmund Koch, Ford Foundation
Professor John A. Lukacs, Chestnut Hill College, Philadelphia
Mr. Robert Mallary, Pratt Institute
Miss Iris Murdoch, Oxford
Mr. Zdzislaw Najder, Warsaw
Professor W. R. Niblett, University of London Institute of Education
Professor Michael Polanyi, Oxford
Professor Edward Pols, Bowdoin College
Professor William H. Poteat, Duke University
Professor F. S. Rothschild, Hadassah Medical Organization, Jerusalem
Professor John R. Silber, University of Texas
Professor Otto von Simson, Free University, Berlin
Professor Lester Singer, New School for Social Research
Professor Charles Taylor, McGill University

HOBBES AND THE MODERN MIND
AN INTRODUCTION

Marjorie Grene

I

In his lectures, *Of Molecules and Men*, Francis Crick celebrates the approaching end of what he calls 'vitalism' and the triumph of a scientific mentality which will envisage all of nature, human and animate as well as non-organic, in terms of the laws governing the behavior of its least parts. He writes:

> Once one has become adjusted to the idea that we are here because we have evolved from simple chemical compounds by a process of natural selection, it is remarkable how many of the problems of the modern world take on a completely new light. It is for this reason that it is important that science in general, and natural selection in particular, should become the basis on which we are to build the new culture. C. P. Snow was quite right when he said there were two cultures. . . . The mistake he made, in my view, was to underestimate the difference between them. The old, or literary culture, which was based originally on Christian values, is clearly dying, whereas the new culture, the scientific one, based on scientific values, is still in an early stage of development, although it is growing with great rapidity. It is not possible to see one's way clearly in the modern world unless one grasps this division between these two cultures and the fact that one is slowly dying and the other, although primitive, is bursting into life.[1]

The physicist W. Heitler on the other hand, interprets the same vision, not as triumph, but as tragedy.

> Belief in a mechanistic universe is a modern superstition. As probably happens in most cases of superstition, the belief is based on a more or less extensive series of correct facts, facts which are subsequently generalized without warrant, and finally so distorted that they become grotesque. . . . A mechanistic living creature presents the grotesque picture of a robot such as is frequently portrayed in comic papers nowadays. The witch superstition has cost innumerable

1

innocent women their lives, in the cruellest fashion. The mechanistic superstition is more dangerous. It leads to a general spiritual and moral drying-up which can easily lead to physical destruction. When once we have got to the stage of seeing in man merely a complex machine, what does it matter if we destroy him?[2]

Both these statements occur, it is true, in 'popular' writings by men of science, but they represent the implicit, and often indeed the explicit, principles of many, both in and out of science. It is this confrontation, and this conflict, with which we are here concerned. Or better, we are concerned with its resolution through the reasoned refutation of the former alternative and the reasoned justification of the latter.

A dogmatic and over-simple materialism like that of Crick can appear as triumph only to those who close their eyes to the simple but irrefutable argument stated as long ago as Plato's *Theaetetus*: that the reduction of knowledge to an enumeration of least particulars denies the possibility of knowledge itself, including this very knowledge. Yet this vision, self-contradictory though it is, has held sway over Western minds recurrently for more than two millennia and increasingly for the past three centuries. It is by now as powerful as it is destructive; and it can be overcome, therefore, only by the fundamental rethinking of our philosophical premises. Such fundamental rethinking is slow and difficult. To contribute in some measure toward its genesis and expression is the aim of the Study Group. The manifesto, the papers in this collection, and, on a broader horizon, the epilogue, should indicate for themselves the direction in which our thoughts are moving. Here I want to put our problem more concretely by taking as a model for the contemporary dilemma a single seventeenth-century thinker, Thomas Hobbes. To be sure, Hobbes was by no means directly, or even indirectly, the most influential thinker of his day; but he was the most revolutionary, and therefore the most prophetic, the most paradigmatic for the dilemma we now confront.

II

It is primarily Hobbes' philosophy of nature with which I shall be concerned. For it is the reigning view of the scientific knowledge of nature, the metaphysical presuppositions of that view and its implications for our conception of man, that have produced the intellectual and moral crisis we are attempting to help resolve. Hobbes' politics, I agree with critics like Tönnies, Brandt, Polin or Watkins in believing, follows rigorously from his conception of nature.[3] The concept of 'endeavor,' and the distinction between the natural and artificial on which it rests, are direct consequences of his rigorous materialism and determinism. In this view, to be sure, I disagree with another authoritative group of Hobbes scholars. But this is not the place to argue this question; it has been exhaustively examined by numerous experts and has, in my view, been settled definitely in Mr. Watkins' recent book, *Hobbes' System of Ideas*.[4] In any case, my exposition, in contrast to these specialized debates, can rest on one of two grounds. The reader may take 'my' Hobbes— or, if he prefers, the Hobbes of any of the aforesaid scholars— as a quasi-historical model, rather as Hobbes himself, like other natural law theorists, took his 'state of nature,' not as historical fact, but as an idealized model of what *would* be, were men's passions given free rein outside civil society. Or he may take this Hobbes—as I join Tönnies, Watkins and others in taking him—as in fact the historical Hobbes. In neither case, I must again emphasize, am I suggesting that the thought of this one man, or of any one man, *caused* our crisis. As I have already said, he was indeed less influential than other, more easily compromising figures. But if the Tönnies-Watkins axis of Hobbes' scholarship is correct, Hobbes alone among seventeenth-century thinkers saw through to where we stand today, and the problems we face are the problems that arise from our intellectual condition as minds dominated by Hobbism, or as the heirs of Hobbism. We may therefore take their Hobbes—or, if, as I believe, they are correct, Hobbes himself—as paradigmatic for our problem.

My task in this introductory essay, then, is twofold. First, I shall enumerate a number of fundamental theses of Hobbesian cosmology. Secondly, I shall point out how fundamental problems at the foundations of a number of disciplines, the problems raised by our first group of contributors, follow from the acceptance of Hobbism; how the philosophical responses of Part II relate to the many-sided dilemma opened up in

3

Part I; and, finally, how the Epilogue suggests a broader range, both of the basic problem and of its possible resolution.

III

As I have opposed, at the start, two interpretations of the modern situation, so I may begin now by confronting Hobbesism and its contrary in Hobbes' own time. Cudworth, in his *True Intellectual System*, was attacking, in his rambling way, 'atheists' as such and, in some places, Hobbes in particular. Hobbes appears to him, and rightly, as the metaphysical leveller *par excellence*. Writing of 'atheists' in general, but clearly attacking Hobbesian nominalism in particular, he writes:

Wherefore they conclude, that there is no such scale or ladder in nature, no such climbing stairs of entity and perfection, one above another, but that the whole universe is one flat and level, it being indeed all nothing but the same uniform matter, under several forms, dresses and disguises; or variegated by diversity of accidental modifications; one of which is that of such beings as have fancy in them, commonly called animals; which are but some of sportful or wanton natures, more trimly artificial and finer *gamaieus*, or pretty toys; but by reason of this fancy they have no higher degree of entity and perfection in them, than is in senseless matter: as they will also be all of them quickly transformed again into other seemingly dull, unthinking and inanimate shapes.[5]

This thesis, the denial of hierarchy in nature and the assertion of a one-levelled ontology, is, indeed, the fundament of Hobbesism, from which all else follows. Not that we could now—or would wish to, if we could—revert to Cudworth's orthodoxly theocentric world. But if we are to find ourselves again at home in a significant universe, we must somehow find, dialectically, a synthesis of what Cudworth asserted and Hobbes denied. If man is to find anew his place in nature, it will not be man as Cudworth knew him, yet somehow man as man, not as a congeries of macromolecules. And even more plainly, Cudworth's God has gone past recall—one can scarcely regret it; yet here, too, as Miss Murdoch argues, it is some analogue of the traditional deity we have to seek, and find, if the fundamental meaninglessness of the Hobbesian world, *our* Hobbesian world, is to be overcome.

4

What, then, is the character of the one-levelled universe so radically opposed to Cudworth's structured cosmos? First, be it noted, it is a universe constructed on the foundation of the contrast between the *natural* and the *artificial*. There can be here no 'higher' and 'lower' in nature. Yet in human life, in what we call culture, in language, custom, institutions, we find nature transformed by man. Whatever else we have done, we have changed the face of the earth: have destroyed much of it, indeed, but have also covered it with novel sounds and sights. To many, these products of human activity, laws, theories, works of art, have seemed higher realities, or the expressions of higher realities, to which we owe allegiance. In a one-levelled world, they can be interpreted only, in contrast to what 'really' is, as artifacts, as what we have *made* in contrast to what naturally exists.

This contrast, between 'nature' and 'convention,' as Hobbes, of course, well knew, goes back to ancient Sophism. Indeed, the *Leviathan* may be seen as the *Republic* minus the Good, with Thrasymachus putting his position rigorously instead of shifting his ground, and with the elaboration by Glaucon and Adeimantus, and the construction of an *artificial* society based on natural need, standing as the final word on justice. Confronted with this contrast, Plato found it necessary, through the difficult indirections of the central books of the *Republic*, to establish a higher, and transcendent, Nature: the Good itself, as criterion for the problems of practice and of knowledge. In relation to this higher and really real nature, the restless, need-driven nature of the phenomenal world is judged, and takes its due, inferior place. Without it, need alone, and in the last analysis, the only 'natural' need, self-assertion, dominates. Our 'standards,' rooted in no higher truth, become conventions only.

The heir of this sophistic view in modern times is the evolutionists' contrast of nature and culture. Thus evolution is either 'natural': the play in endless statistical diversity of survival-for-survival-for-survival, the self-perpetuation of genetic combinations in a war of all against all at the level of gene-environment relations; or 'cultural': the self-perpetuation of man-made artifacts, whose natural end, however powerful their apparent authority within a given society, is again survival alone. Not the good life, but life, is the goal of living.

Now this is precisely the contrast on which Hobbesism essentially relies. Both Hobbes' theory of language (and of knowledge) and his theory of society rest explicitly on this foundation. 'What,' Hobbes wrote to Descartes in protest against the innatism of the Third Meditation, 'if reason be nothing but the stringing of names together?'[6] In a one-level universe, our discovery of the rational order inherent in speech appears, not as the mediator of a cosmic Logos, but as a pure *invention*. The conventional element in language takes over altogether and speech becomes an aggregate of insignificant noises or meaningless marks, used in order to tie together bundles of happenings— and in themselves a trivial series of happenings, too. Hempel has described mathematics as a giant juice-extractor: it conveys no information about the world, but serves to press out of the raw material we put in the juice we want to get out.[7] Hobbes, enamored from middle age of Euclid, saw all language as having this mathematical character. Word-borne thought, for him, is *calculation*, and deductive systems, the machines (computers, we would say) that we use to help us calculate. 'For words are wise men's counters, they do but reckon by them. . . .'[8]

It may be instructive to contrast here, very briefly, the reliance by Hobbes and Descartes on mathematics as the paradigm of scientific knowledge. Hobbes, as we all know, was a poor mathematician and Descartes a great one. Yet Hobbes' conception of mathematics and its uses comes closer to the conceptions now current than does Descartes'. For Hobbes had seen through to the positive outcome of the scientific revolution. It is not so much that, in Galilean phrase, nature is written in the mathematical language, as it is that mathematics, in itself blind and empty, is the only instrument through which we can suitably manipulate the happenings of nature as our senses record them. Descartes, on the other hand, was still sufficiently confident of the *real* power of mind to affirm the direct ontological import of geometrical thought within the bounds of clarity and distinctness. To be sure, this ontological claim had to be supported by God's unwillingness to deceive and by the vestige of hierarchy needed to enclose on the one hand such a non-deceiving God and on the other such well-coordinated finite natures as thinking minds and extended bodies. But within these boundaries Descartes' emphasis is on the active power of the mind to

6

know, and, in knowing, to submit itself to the being whose luminousness is its judge. Hobbes, on the other hand, transforms this action into making, and this power into the sheer unrestricted doing of the artificer. Descartes, so far as I know, shows no interest anywhere in the problem of universals; but his theory of innate ideas marks him as, implicitly, a believer in a realist theory. For the mind has innately the power, not of inventing juice-extractors, but of acquiring from within itself insight into truth. In contrast, Hobbes is the arch-nominalist. Hobbesian systems have to have truth fed into them by means of sensory input; in themselves they are merely machines for the efficient storage and translation of such information. For Descartes, theory is still vision; for Hobbes, construction.

Tillich, in *The Courage to Be,* calls nominalism a forerunner of existentialism, in that it paves the way for that mood of anxiety, the anxiety of meaninglessness, which characterizes the thinking of our time.[9] This relation clearly obtains for Hobbesian nominalism: words are wholly deprived of meaning and so, as we shall see, is the nature through which, by their aid, we thread our way. Alienation is already complete and dread lies in wait.

One qualification should perhaps be made of this sweeping statement of Hobbes' conventionalism with respect to language. Frithiof Brandt, in his illuminating book on Hobbes' mechanical system, suggests that Hobbes, after his rash remark to Descartes about reasoning, came to realize that reason is not only 'stringing of names together,' that *concepts* must be correlated to words if the latter are effectively to serve our purposes.[10] Concepts, however, for Hobbes, are *images,* 'phantasms' of sense. They are themselves both particular and meaningless. They add neither generality nor significance to the empty calculations of discourse. Indeed, as we shall see shortly, it is precisely the supplementation of verbal calculation by reference to sensation that gives Hobbes, three centuries in advance of Schlick or Carnap, the model in which the chief contemporary conception of science has come to rest.

I need not stop here to elaborate the way in which Hobbes' politics also rests on the opposition of nature and art. Warrender's subtleties notwithstanding, I need only point to the justification of *Leviathan*'s title: 'Nature . . . is by the art of man, as

7

in many other things, so in this also imitated that it can make an artificial animal.'[11] This is, again, the *Republic* without the Good. The state has been devised to satisfy natural need, i.e. survival, by evading natural peril, i.e. the sudden death which threatens each from each other's fear, vanity, and greed. One-levelled nature controlled through artifice: this is still the one opposition which Hobbesian, like positivist, thought permits.

In such a one-levelled universe, what is knowledge? There are two possibilities, which Hobbes unites in a theory exactly equivalent to the now orthodox conception of scientific explanation. The single source of information about the world is sense; everything we know is fed in by the senses and retained, as they 'decay,' in memory. This is 'absolute knowledge.' Science, however, is hypothetical: it imposes axiom systems, themselves empty, upon the sensory base. Thus we have what has come to be known as the hypothetico-deductive model of explanation. Theoretical constructs, like the atomic theory, for instance, are superimposed upon the sensory data expressed in protocol reports. In themselves, theories make no claim to truth; they are vehicles for calculation, for retroductive summary and storage of sensory input and predictive claims to future sensible results. The *explanans* is a statement with empirical content from which the *explanandum*, and its observable derivatives, can be deduced. True, Hempel and Oppenheim, in their now classic paper on explanation, timidly (and in their own terms, unintelligibly) assert: 'the *explanans* must be true.'[12] But this is clearly an embarrassment, as Hempel testifies in 'The Theoretician's Dilemma.'[13] What hypotheses *should* be, really, on this view, are not claims to truth, since claims are personal and there are no persons in this universe, but paper-and-ink computers, machines for turning sensory input into sensory output.

There has seemed, however, to be a difficulty in this theory, a difficulty similar to that which haunts Hempel's 'dilemma.' Scientists construct, not arbitrary hypotheses, but hypotheses of one particular variety. Epicurus, holding a similar methodology, had insisted on multiple causality; an indefinite number of explanations, he argued, might be the premises from which a given phenomenon could be shown to follow and hence, in his sense, its causes. And until one hypothesis rather than another produces observable consequences, whether in confirmation or disconfirmation, this equivocity may indeed obtain. Yet scientific explanation does follow certain conceptual lines and not others. Why? Or why not? In particular, Hobbes, and his twentieth-century positivist heirs, construct systems, the basic terms of which refer in one guise or another to *matter in motion.* For Hobbes, the proper vocabulary was still the traditional one of 'substance,' but substances for him were simply bodies, and motions and countermotions the accidents of bodies. For twentieth-century physicalists, the proper concepts boil down to events, not substances, space–time coordinates and ultimately the particles (or waves) specified in terms of such coordinates. Why only this materialist system and no other? Hobbes' answer is essentially the Epicurean one: no other construct, no other verbal machinery, would fail to contradict the data of sense.

Admittedly, Hobbes does not state this principle explicitly in his mature work on nature, the *De Corpore.* Here he simply constructs his hypothetical corporeal language, then applies it in what he calls 'Physics' to the phenomena of sense. Nonmaterialist language, of spirits or ghosts or what-not, he dismisses as self-contradictory. In the earlier *Elements of Law,* however, he makes this reason for his materialism somewhat more explicit.[14] Materialism is the only successful system because it is the only system not contradicted by sense. Despite, or even because of, the purely conventional character of science, its theories are restricted to the terms of matter in motion.

To support this restriction, of course, sense itself, which provides the content of our otherwise empty system, must be shown to be reducible to motion. So, Hobbes points out, if you strike your eyeball sharply, you see light. Seeing is the consequence of, indeed, *is,* a motion. And similarly for the other senses. This link given, he can construct a detailed theory of the causes of sensory phantasms in the rigorous terms of matter in motion. And, again, the translation of his now antique language, whether into terms of the electronics of brain action or the chemistry of enzyme specificity, has made no single jot of difference in the underlying philosophic view. All explanation is hypothetico-deductive, and the only meaningful hypotheses are such as reduce all phenomena to material or, in modern language, to physico-chemical terms.

9

Such reduction, moreover, means reduction of wholes to parts. Explanation is hypothetico-deductive, materialistic and *particulate*. True, Hobbes is not in the Democritean sense an atomist. Hobbesian bodies are not atoms and the space which he postulates at the beginning of his system is an imaginary, not a real, void. His interest is different and the turn of his atomism differs accordingly. In sympathy with Descartes and Galileo, he is trying to formulate rules for a mechanical nature, where the new discoveries about motion will prevail. He is willing to take 'substance' simply as equivalent to 'body,' and body as 'filling some part of space.' But the *motions* of bodies—and that is what all alterations in nature come down to—are analyzed in terms of their least unit: *conatus* or endeavor. All mutation is motion and motion is particulate: the continuity of motion is understood as a summation of minute motions. Nor, since all change is motion, can there be wholes not reducible to such least parts. There are only bodies moving, and the analysis of their motion resolves these into parts. Hobbes is an atomist, not a Democritean, but an atomist of motion.

Where we find wholes, therefore, they are either apparent or artificial. In nature the real accidents of bodies are their motions, which are, as we have just seen, analyzable into their least parts. The colors and sounds that seem to belong to bodies, however, are phantasms in the subject, and explicable in terms of the corporeal motions that cause them. Although he does not state it as such, Hobbes embraces wholeheartedly the Galilean-Democritean distinction between the mechanical qualities of bodies and those appearances which, in Galileo's terms, 'would be mere names if the living creature were removed.'[15]

Apart from these 'phantasms,' which are mere seemings, moreover, such wholes as we have to deal with in our experience are merely *made* wholes. They are language systems which constitute sciences when applied to sense, or legal systems invented to assuage our fears and satisfy our needs: needs and fears which, in turn, reduce to a sum of 'endeavors' or least motions of our bodies to and from other bodies which have in turn caused such minute internal movements.

All motion, moreover, is determinate. There are no tendencies, no aims. In this again Hobbes shares the Galilean and Cartesian resolve to eliminate final causes from nature, and

10

equally the Spinozian resolve to eliminate ends from the explanation of human nature. There are in nature only causes, no reasons. Appetites and aversions which appear to be directed to and from objects are summations of endeavors outward in reaction to endeavors inward from the thing. There are simply local motions, and, when these conflict, the resolution of the conflict has to be devised by artifice through the contractual institution of civil society, which our drives have driven us to contrive for the avoidance of our own destruction. Modern explanation, Stocks has remarked, neglects Aristotelian formal and final causes in favor of material and efficient causes alone, agencies which in Aristotelian terms would be intelligible only as correlates of their respective partners, form and end.[16] This program is rigorously carried through in Hobbesian mechanism: as wholes are explained through reduction to their parts, so ends through their reduction to a series of undirected motions. All 'choices' are in effect Hobson's choices, 'the choice of the junior at meat.'

It is scarcely necessary to add that such determination also abolishes 'oughts' and, like Hume's argument a century later, subordinates reason to passion. Where there are no reasons but only causes, there can be, except by the devices of civil society—themselves motivated by the causes of fear, greed and honor—no obligations. The laws of nature for Hobbes, as Watkins convincingly argues, are not 'laws' in a normative or ethical sense; rather they are of the nature of a physician's prescription.[17] The doctor assumes we want to get well and prescribes means. First comes fear of pain or death and/or desire of health, then consultation, prescription, submission to a medical regimen, and, then, it is to be hoped, recovery. Similarly, self-interest, using the calculation that is 'reason,' reckons out the need to keep the peace and submits.

One final, and fundamental, point. All this, for Hobbes as for modern empiricist philosophies, is meant to be no metaphysic but a non-speculative system, arbitrary in itself, but tied down through control by sense. Yet Hobbes' materialism is only apparently non-metaphysical or non-dogmatic. As we have seen, the link between hypothetical system and observation depends on the thesis that all sense is motion. This is the single and indispensable link that ties theoretical calculation to sensory

11

in- and output. But the thesis that all sense is motion is only fleetingly argued for in the *Elements*, in the *De Corpore* not at all. Nor is it, either logically or empirically, self-evident. Logically, it would be in Hobbesian terms a mere matter of convention so to decree. Empirically, it must be admitted there is no phenomenal uniqueness about motion, let alone about 'matter' in motion. As Berkeley has conclusively shown, primary and secondary qualities are, as phenomena, exactly on a par. There is no unique access here to motion, nor to some underlying bodies whose real accident it is to move. What remains, then, as the ground for Hobbes' essential thesis is pure metaphysical faith. Only the unquestioned belief that all mutation *must* be motion makes him so sure that in fact it is so, and that therefore his free-floating system of 'reason' can, alone of all hypothetical systems, be tied down to its roots in sense. This is, in other words, like modern empiricisms, a singularly narrow metaphysic disguised as anti-metaphysical. It is indeed the same metaphysic concealed under the same methodological disguise that we meet again, not only in logical positivism, but in the central state materialism of Smart or Armstrong, or in the computer-ridden speculations of writers like Putnam or Scriven. And it is the pervasive influence of this particulate, materialist, one-levelled ontology that generates in one way or another the problems our participants confront.

IV

Hobbesism, as I have sketched its principal tenets, is the framework within which our problems, and against which our adumbrations of solutions to these problems, arise. Let me indicate briefly how, in my view, the theme of our contributions bears on this central issue.

Professor Wigner's paper, first, exhibits physics and psychology in a paradoxical interrelation generated by the phenomenalism of the one and the would-be materialism of the other. The two supports of Hobbesian science, sense and a corporeal language, have split apart and confront one another in apparent contradiction. Physics, far from its earlier confidence in the explanatory power of the 'corpuscular philosophy,' has, with the dissolution of the classical atom, dissolved its meta-

12

physical claims and puts its trust in 'observation' as such. In other words, it reports, not the character of an inferred material world, but observations pure and simple: Hobbesian phantasms of sense. But their only 'reality' is our awareness of them: the world is turned back into a non-interpreting, self-enclosed consciousness. Psychology, on the other hand, still caught in a physicalist metaphysic which physics has long ago transcended, would eliminate 'consciousness' as mere seeming and translate the data, Hobbes-like, into a language of matter-in-motion. On the one hand, 'mechanical' nature has dissolved into the reporting of the merely sensed, and on the other, sensing is held, if in more sophisticated language than Hobbes', but still in the same spirit, to be no more than motion. We have, on the one hand, the science of motion reduced to consciousness and, on the other, the science of consciousness allegedly reduced to motion. Only a new reading of mind and nature, and of mind in nature, can resolve this impasse.

At the same time, contemporary physics, with its esoteric concept of 'observation,' has come far from the assessment of phenomena themselves in their full experienced lawfulness. For the 'real' qualities of the 'real' world were from the start conceived to be, as Hobbes, too, conceived them, those qualities alone which seem to associate themselves most intimately with matter in motion: size, shape, position and motion itself. All else—color, sound, smell and the qualities grasped by touch—were phantasms only. Accordingly, the 'observation' praised by Wigner as physics' strict base is in fact the refined and indirect reporting of 'primary qualities,' in particular of position and local motion, in the remotest abstraction from our ordinary experience. In this, some people have argued, physics has unnecessarily narrowed its own field and so has exacerbated the conflict that issues in Wigner's paradox. In favor of the primacy of measurable motion, sense has been reduced to the thin thread of observation most readily amenable to such interpretation, and in turn the discipline that claims to interpret sense has been confined by this over-abstract aim of physics, understood as the aim of science-*qua*-science.

One of the early protests against the implications of this over-abstraction came from Goethe, who argued in his *Farbenlehre*, not, indeed, that Newton's treatment of color was mistaken,

13

but that it was incomplete. In our second essay, Mr. Holtsmark expounds Goethe's main thesis: his insistence on the scientific lawfulness of color phenomena themselves, and shows how, if we take Goethe's argument seriously, we may lessen the gap between science and science, between nature and mind. Nor is this a piece of historical antiquarianism. There is important contemporary work—Holtsmark refers to the experiments of E. B. Land on color vision—which is strikingly reminiscent of Goethe's approach.

What has this to do with our Hobbesian model? The *Farbenlehre* shows us, in its direct investigation of color phenomena, how the perceived world in its very surface as perceived is impoverished by the cramped reduction to motion insisted on by Hobbesian thought. That reduction, moreover, is, as we have seen, a reduction of natural wholes to their particulars. This, too, Goethe's argument, and Holtsmark's, opposes, or at least urges us to supplement:

> Da hat man die Teile in seiner Hand;
> Fehlt leider nur das geistige Band.

The gestalt-like aspect of phenomena, Holtsmark argues, is worth studying, both by physicists and psychologists, for its intrinsic interest. And it is of crucial importance, further, for philosophers seeking, as members of the study group are seeking, approaches to a renewed and more adequate assessment of the phenomena that confront us over the whole range of the biological and the human. Except in his concluding remarks, Mr. Holtsmark has not explicitly emphasized this in his paper; but, if we put his exposition, for example, alongside Polanyi's contribution on the social sciences and Koch's on value properties, we can see opening up a perspective essential to our common endeavor: that is, the renewed reflection on the relation between facts and values.

The conception of value-free, absolutely 'factual' facts is tied essentially to the conception of the world as particulate: to Wittgenstein's early theme, 'the world is all that is the case,' and the (vain) search for atomic sentences to pin down the atomic facts that compose it. Contexts, on the contrary, entail appraisal, that is, evaluation; only the isolated datum can be value-free. When, therefore, we abandon the search for the

14

MARJORIE GRENE

particulate as the unique and universal maxim of scientific inquiry, we abandon by the same token the illusion that we can find facts-in-themselves shining in splendid isolation in a heaven of objectivity. Wholes, once we abandon our Hobbesian faith, can indeed be as natural and real as parts. But, on the one hand, they can be found to be so only within the context of our evaluation of them; and on the other, the acknowledgment of their existence is at the same time the acknowledgment of their significance, of the existence, that is, of *real*, not merely conventional or purely enacted, values. Thus, facts become evaluative and values, we discover, can be discovered as well as made.

Such a change in perspective, even within the usual subject-matter of physics, or at least at the confluence of physics and psychology, augurs well for the continuity we hope to restore between a series of levels or dimensions of reality and between a series of disciplines, from exact science through the behavioral sciences to the arts, in which we seek to describe, explain and understand those distinctive yet interpenetrating realities. Only through such an effort, finally, can we hope to conquer Hobbesian determinism and restore 'oughts' to an authoritative role with respect to the brute 'facts' of aversion and appetite. (This last point: it is sometimes alleged that the indeterminacy principle of modern physics has already accomplished: it should be noted, however, that statistical laws are just as necessary in their import, and therefore in their reduction of responsible action to a mere byplay of physical causality, as are the laws of classical mechanics.[18] As Wigner's essay shows, quantum-theoretical epistemology only displays dramatically the paradox of Hobbesism, but does not begin to answer it.)

If, as Holtsmark suggests, the assessment of perceived wholes is sometimes appropriate even in areas studied by physics, or at the meeting-point of physics and psychology, so much the more is this the case for the biological sciences. There have been recurrent attempts to interpret organic phenomena in what we are here calling Hobbesian terms. One such was the 'mechanism' of the early twentieth century; neo-Darwinism, though conveniently embedded in protective ambiguities, seems to me on the whole to be another. But the most conspicuous biological Hobbesism of the moment is the claim, eloquently

15

voiced by Crick in the lectures referred to above, that biology is on its way to becoming wholly a science of macromolecules. True, the biologist need not concern himself, for the most part, with the 'nuclear jungle,' but neither need he concern himself any longer, on this view, with the organisms of which his discipline traditionally treats. His interest is wholly in the particulars of which these 'entities' are compounded, the parts which explicitly determine their behavior and so exhaustively specify their nature. To take them as 'wholes' is either a convenient fiction or the mere vestige of a superstitious age. What we have 'really' is either a series of determinate events or an aggregate of them amounting to a crowd of random motions governed by statistical laws. Nor of course, *a fortiori*, can anything in the way of trends, of final causes, be admitted to their nature. Mechanical cause and chance are all in all.

This proclamation of Hobbesism rampant is challenged in different directions by both our biological contributors. Barry Commoner examines the alleged reduction of biological systems to their parts with respect to the practice and theory of molecular biology. C. F. A. Pantin compares the abstractive and restricted character of the exact sciences with the unrestricted nature of biology. Biological science, he argues (in common with some other disciplines, like geology), needs to draw for its methods and its insights on many fields and areas of information as well as on many and disparate techniques. It has not, as on the fashionable view it ought to have, the monolithic character of Hobbesian science. This property of non-restrictiveness is linked, moreover, to the richness of the organism-environment relation which the biological observer-experimenter takes for granted—and has to take for granted—both for his subject and himself. Thus, while Commoner is challenging the particulate materialism of a Hobbesist world view in the face of the existence of real systems that can be assessed and, indeed, analyzed, as such, Pantin is challenging the Hobbesian conception of sensory input as a series of discrete givens which carry in themselves no resonance either to other givens or to the larger contexts out of which they appear and whose organized character they express.

These papers, like Wigner's and Holtsmark's, raise fundamental questions: they indicate points at which the Hobbesian paradigm reaches the limits of its powers and show us something about the lines along which, in the face of such limitations, we may seek to articulate a more tolerable position. To a philosophical reader, however, they also present difficulties which have yet to be thought through and overcome. Commoner would insist that everywhere science has to take account of naturally existent wholes. If DNA experiments *in vitro* have not yet solved the 'riddle of life,' as proponents of the 'central dogma' have insisted, neither, he argues, do micro-theories in physics suffice in all cases to save the phenomena they are alleged to save. In support of this allegation, he cites Bardeen's report on microscopic versus phenomenological theories of superconductivity, where, he argues, only reference to the whole has permitted physicists to explain phenomena which otherwise resisted explanation. Bardeen's paper, however, expresses unambiguously the confidence that a micro-explanation in this field will ultimately obtain: the same confidence which programmatic microbiologies express for the phenomena of life.[19] There may indeed be a revolution in physics to come which will justify Commoner's general position.[20] As far as his evidence goes, however, he seems to be arguing what no biologists, however 'micro,' would deny, that *heuristically* we must often examine the whole in order to find the part-determined explanation we seek. This says nothing about the *ontological* character of whole-part relations, within or beyond the range of living nature. The problem of the structure of biological as against physico-chemical explanation, moreover, Commoner treats sometimes as an empirical one: DNA theorists have not yet achieved the reduction they seek—and sometimes as a conceptual one: organic phenomena are of such a structure as essentially to resist such reduction. Thus the question whether the phenomena of the non-organic world demand on principle 'holistic' explanation, not only on the boundary between physics and psychology, as in Holtsmark's example, but more comprehensively, and the question whether biological phenomena are in this respect like or unlike those of physics, both remain open. Yet they are raised in Commoner's paper in a ringing challenge which demands further exposition and resolution, a resolution for which some of the papers in Part II, notably those of Pols and Polanyi, indicate a possible direction.

Pantin's paper suggests the need to open doors in another,

epistemological direction. As a scientist sensitive to his philosophical environment, he notes the apparent adequacy of a sense-datum theory—i.e. at bottom a Hobbesian theory of phantasms—for the procedures of physics as classical and quantum physicists have understood them, but doubts its adequacy to account for ongoing organism-environment relations as the biologist both encounters and exemplifies them. Again, a reform of the theory of perception which would meet these needs is already demanded by Wigner's paradox as well as by such counter-examples as Holtsmark's. And again, the adumbration of such a reform is at hand in the theory of tacit knowing introduced in Polanyi's essay on 'The Structure of Consciousness.' So we have here another example of the impasse to which, vis-à-vis biological thinking, the Hobbesian paradigm leads, and in our 'philosophical responses' an indication of a road by which the impasse may be avoided and a solider conceptual foundation laid for the many-dimensional research in which biologists in fact have been and continue to be engaged.

The recognition of wholes not explicitly resoluble into parts, the recognition that evaluation plays an essential role in scientific judgment, the reasoned rejection of a one-levelled determinist view of nature: all these are demanded by the difficulties in the interpretation of nature, living and non-living, to which Hobbesism has condemned us. If these problems confront us in other fields, so much the more urgently do they appear in what was once the science of 'consciousness,' psychology. The dilemma of psychology, Sigmund Koch argues, is still behaviorism. To its classical form a neo- and a neo-neo-version have succeeded; in any of these disguises it is still the Hobbesian framework that hampers the psychologist's experimental design and impoverishes his conclusions. Koch's chief concern is with motivational theory: he wishes to resist the exclusive use of means-end concepts in analyzing human behavior, and to allow instead an approach in terms of 'value properties' which will permit the analysis of much meaningful, yet non-motivated, behavior. And it is precisely the theme of the value-borne character of human 'facts' that emerges here, more explicitly than in Holtsmark's or in the biological papers, as a central *leitmotif* of our discussions. Koch writes:

18

Differentiated value events . . . are omnipresent in psychological function. If fact and value are ontologically disparate or in some sense separated 'realms' or aspects of the universe, I do not know what the *psychological* evidence could be. I do not in fact see how one can conceive of any such monster as an axiologically neutral fact, or, for that matter, a factually neutral value.[21]

Nor does this rethinking of the alleged fact-value dichotomy announce a program of 'subjectivism'; it heralds a liberation from the cramped S-R model of Hobbesian psychology to more varied and also more precise insights. Indeed, it makes a broad spectrum of significant behaviors accessible for the first time, in Koch's view, to detailed experimental study.

True, there seems at first sight a paradox here in relation to our thesis of the Hobbesist paradigm. Motivational theory interprets all behaviors as means-end directed; Koch would free them from this confinement and recognize that we perform many actions, not in order to satisfy drive X or Y, but just because we like to. Hobbesian analysis, on the other hand, itself denies means-end relations or final causes altogether and so seems to resist the very type of explanation that Koch too is rejecting. But the contradiction is only apparent. In fact, Koch is reinstating a concept of *intrinsic* meaning in psychology, where previously all meaning had to be extrinsic. And the latter is precisely the Hobbesian theme: both phantasms—the appearances of motion—and motions themselves are inherently meaningless; significance (a property of wholes, not of particulars) is in itself artificial, decreed from outside and unreal. It is thus once more the mechanical and particulate determinism of a Hobbesist cosmology that forces motivational theory into its monolithic mold. 'Drives' are only apparently end-directed; in fact they operate to push behavior from behind. Thus there must be for every action a drive which triggers it, a movement from the environment in reaction to which it moves, in endeavor after endeavor, until, with the force of the original stimulus exhausted, another takes its place. And this occurs not because there are real final causes, but because there are none; the indefinite accumulation of meaningless small movements is what action amounts to. Behavior, seen truly, is but the macroscopic appearance of microscopic particulars simply located in space as well as in time. In other words, we have put first in man,

19

as Hobbes did, 'a restless desire of power after power, which ceaseth only in death.'[22] So natural tendencies, with their own natural rhythms, are turned around into drive-behavior sequences in which the single unidimensional course of rectilinear motion is maintained. (Similarly, as Charles Taylor has shown, learning theorists quickly suppressed the scandal of Thorndike's 'law of effect' by interpreting it in traditional S-R terms.)[23] And it is Koch's rejection of the apparent means-end straightjacket that liberates psychological subject-matter once and for all for the inspection and analysis of naturally directed behaviors: of finite sequences which are self-sought and self-contained.

If, moreover, Hobbesism has led experimental psychology into emptiness and paradox, so much the more glaringly has this occurred in the development of social science. The absurdity of a purely 'factual' account of social and political life is the theme of Polanyi's essay, 'The Message of the Hungarian Revolution.' Hobbes identified his matter-in-motion language with science; arbitrary in itself, it was nevertheless the only system which, confronted with sensory data, would prove non-contradictory. Social scientists have loudly and firmly accepted this view. By it, however, persons as naturally existing entities or nexus of natural process are abolished; both the scientist and his subject become aggregates of mechanical responses to mechanical stimuli. A distinction between 'drive-satisfaction' and a claim to rightness or truth becomes an absurdity; the latter is at best a trivial grace note to a rigorous determinist theme. Thus, Polanyi has written elsewhere:

... it has now turned out that modern scientism fetters thought as cruelly as ever the churches had done. It offers no scope for our most vital beliefs and it forces us to disguise them in farcically inadequate terms. Ideologies framed in these terms have enlisted man's highest aspirations in the service of soul-destroying tyrannies.[24]

And again, whether the laws discovered by this method be classically mechanical or statistical in formulation makes no difference to the rigor with which they reject such claims to truth or right. Nor does the injection of 'value premises' into the starting-point of a sociological or political investigation rectify this rejection. For such addenda can always be retranslated into terms of 'appetite and aversion.' A much more comprehensive reform is needed here, a recognition, once more, of the claim to truth—moral as well as factual truth—entailed in responsible action. But until the active nature of being as persons, and of social existence as the constitutive ground of such being, is once more understood and accredited as the framework of behavior, such recognition is impossible. I am not, it must be emphasized, suggesting that what is involved here is a plea for an 'organic' theory of society, nor, in the interpretation of individual action, a reinstatement of 'dualism,' 'vitalism,' or any such rightly defunct 'ism.' What is needed is a new fundamental theory of persons and of societies; for there are indeed no persons apart from society: in this Hobbes' insight is correct. Some existential accounts of 'being-in-the-world' come close to such a view, and Polanyi's work is, indeed, strikingly convergent with these; but what we need is to make such descriptions sufficiently general and sufficiently well grounded ontologically to implement the inclusion, rather than, in existentialist fashion, the exclusion of science, both its heuristic and explanatory procedures and the diverse characters of its many subject-matters, among which communities of persons may then take their due place.

Polanyi's essay speaks plainly for itself; but it may help to show its place in our common effort and also to state more comprehensively than I have done so far the broader purpose of our enterprise if I quote here a programmatic statement from Polanyi's major philosophical work. Insisting that all search for knowledge, however 'objective,' takes place within a framework of intellectual commitment, he shows how this perspective serves to restore the possibility of recognizing our own activities and our own world, cultural as well as natural, as an ordered continuum of meaningful contexts rather than, in Hobbesian fashion, as an aggregate of blind appetitive movements. Near the close of Part III of *Personal Knowledge*, he writes:

Within the framework of commitment, to say that a sentence is true is to authorize its assertion. Truth becomes the rightness of an action; and the verification of a statement is transposed into giving reasons for deciding to accept it, though these reasons will never be wholly specifiable. We must commit each moment of our lives irrevocably on grounds which, if time could be suspended, would invariably prove inadequate; but our total responsibility for disposing of ourselves makes these objectively inadequate grounds compelling.

Truth conceived as the rightness of an action allows for any degree of personal participation in knowing what is being known. Remember the panorama of these participations. Our heuristic self-giving is invariably impassioned: its guide to reality is intellectual beauty. Mathematical physics assimilates experience to beautiful systems of indeterminate bearing. Its application to experience may be strictly predictive within certain not strictly definable conditions. Alternatively, it may merely express a numerically graded expectation of chances; or provide only—as in crystallography—a system of perfect order by which objects can be illuminatingly classified and appraised. Pure mathematics attentuates empirical references to mere hints within a system of conceptions and operations constructed in the light of the intellectual beauty of the system. The act of acceptance becomes here entirely dedicatory. The joy of grasping mathematics induces the mind to expand into an ever deeper understanding of it and to live henceforth in active preoccupation with its problems.

Moving further in this direction, we enter on the domain of the arts. Once truth is equated with the rightness of mental acceptance, the transition from science to the arts is gradual. Authentic feeling and authentic experience jointly guide all intellectual achievements; so that from observing scientific facts within a rigid theoretical framework we can move by degrees towards dwelling within a harmonious framework of colours, of sounds or imagery, which merely recall objects and echo emotions experienced before. As we pass thus from verification to validation and rely increasingly on internal rather than external evidence, the structure of commitment remains unchanged but its depth becomes greater. The existential changes accepted by acquiring familiarity with new forms of art are more comprehensive than those involved in getting to know a new scientific theory.

A parallel movement takes place (as we shall see) in passing from the relatively impersonal observation of inanimate objects to the understanding of living beings and the appreciation of originality and responsibility in other persons. These two movements are combined in the transition from the relatively objective study of things to the writing of history and the critical study of art.

The growth of the modern mind within these great articulate systems is secured by the cultural institutions of society. A complex social lore can be transmitted and developed only by a vast array of specialists. Their leadership evokes some measure of participation

22

in their thought and feeling by all members of society. The civic culture of society is even more tightly woven into the structure of society. The laws and the morality of a society compel its members to live within their framework. A society which accepts this position in relation to thought is committed as a whole to the standards by which thought is currently accepted in it as valid. My analysis of commitment is itself a profession of faith addressed to such a society by one of its members, who wishes to safeguard its continued existence, by making it realize and resolutely sustain its own commitment, with all its hopes and infinite hazards.[25]

Before we turn to adumbrations, in our second group of papers, of a philosophy which would implement this program, there is one more discipline to be included, whose problems lie especially close to the heart of our enterprise. If social scientists have labored massively at the reduction of their subject-matter to pure facticity, legal theory has, as John Silber argues, only partly succeeded in its efforts to become a Hobbesian discipline. It has tried, indeed—not only, of course, in positivist systems like that of Kelsen, but, in judicial practice, in the attempt to interpret acts as bead-like units on a string of moments and to hold the agent responsible only in so far as we discover him openly and publicly committing such momentary acts. Now, of course, as Silber admits, the law has also had good reasons of social justice for such an emphasis: we would not wish to hold men responsible, as totalitarianism does, for their secret thoughts, nor, as more static societies did, for the class to which they happen to be born. Yet what men do does express what they are. Indeed, it is for what they are, in the sense of their characters, for what they tend to do rather than for their momentary actions, that we most often and most systematically bestow moral praise and blame. To try to single out isolated moments of action as the sole sources of responsibility, Silber argues, therefore produces anomalies and inconsistencies which only a profoundly different theory of responsible action could correct. He suggests what he calls a continuum theory, ranging from responsibility for one's *being*— legally less culpable but still to be taken into account, by ethics if not by criminal law—to the sharp and clear responsibility for one's *doing*, which, however, is grounded in and carried by the being which one also responsibly accepts as one's own.

23

V

Silber's essay, like Polanyi's, carries us forward to the more general philosophical perspectives of Part II, where in a number of directions our contributors are suggesting lines of philosophical inquiry which might help us to resolve the epistemological and ethical problems that we have found arising, out of the implications of the Hobbesian model, at the foundations of physics, biology, psychology, and social science.

I need not dwell in any detail on the contributions of these papers, for the problems of which they treat have been raised in sufficient number and clarity in Part I. Professor Stallknecht, in an essay relying fundamentally on Whitehead's program for metaphysics, sets the broadest philosophical range of our endeavor. Miss Murdoch stresses the moral aspect of our dilemma, and suggests that, having lost our natural sense of good, we may look for guidance to art and especially to great art. For here we confront realities to which we must say we owe submission, and so exhibit aspects of human experience which the Hobbesian conception of science and of ourselves as known by science cannot envisage. Miss Murdoch in large measure accepts, indeed, a Hobbesian view of man's private, appetitive nature: our lives, left to themselves, are solitary, nasty, and brutish, even if no longer short. But, she argues, there must be something more, and something not merely conventional, to support what is *not* Hobbesian in our experience: the experience of a good, indeed of a *summum bonum*, which is not, as for Hobbes, either the flat assertion of desire or the avoidance of a *summum malum*. Professor Bossart, looking at some contemporary methods in philosophy, suggests a road, in a 'phenomenology of freedom,' which may allow us to articulate such hopes as those which Miss Murdoch has expressed. Finally, Pols and Polanyi seek to apply such a method, or one very close to it, in the philosophical analysis of a special, and central, problem: that of personhood. Pols attacks the problem as an ontologist, and reintroduces in a new context the concept of levels of reality which, as Cudworth foresaw, Hobbesism had banished from nature, to the exclusion of any adequate concept of person or of agency. Polanyi outlines in his theory of tacit knowing an epistemology which enables us to restore such a

24

hierarchical ontology, and shows how this epistemology can be applied for the reinterpretation of consciousness. With this we come full circle to the explicit resolution of the paradox set by Wigner: physics becomes the activity of persons, not the reportage of isolated and unmeaning data, and psychology the discipline in which the many-levelled, deeply evaluative activities of persons can in turn be studied.

VI

The questions raised in Part I stem chiefly from problems arising at the foundations of scientific disciplines. They issue, as our title suggests, in a reexamination of the relation of values, and of responsible agency, to knowledge. At the same time, the aesthetic aspect of experience has been referred to in a number of contexts. It is indirectly involved in Goethe's concern with the phenomenal aspect of color perception; it is a central theme of Koch's program for the study of value properties; it becomes, in terms of Polanyi's theory of commitment, continuous with the subject-matter of psychology and social science; and, in Miss Murdoch's paper, the aesthetic functions as our reentry to knowledge of the good. Indeed, in answer to Crick's prediction of the dominion of a wholly scientific culture, we would reply that only a reinstitution of knowledge *within* an evaluative frame can meaningfully repair the breach of our two worlds, and that within such a frame art holds a conspicuous place. For the aesthetic is, as Peirce insisted, the dominant and ultimate evaluative dimension. Much of the study group's discussion in its two meetings was directed to problems of art, of the nature of artistic discovery, of art and truth, of the role of the artist in society. Unfortunately, however, there were not enough formal papers on such subjects to permit their inclusion here; nor can we within the bounds of one volume cover the whole scope of our concerns. Both for its own excellence, however, and for the glimpse it gives of this broadest cultural horizon of our program, I have included in the form of an epilogue Elizabeth Sewell's poem, 'Cosmos and Kingdom.'

For Hobbesians, art must be amusement merely; it falls plainly within the class of the artificial. It can be shown, of

25

course, that even artifacts display the kind of many-levelled structure Hobbes denied; this is clear from the dichotomy in engineering between the physico-chemical analysis of a machine and the principles which specify its function. The end of such construction, however, is still conceived by many as Hobbesian; all making, art, engineering, even the making of pure science, has motion, power, sheer doing and self-assertion as its sole end. Against such restless appetitive striving, common to man and beast, we can set, in Hobbesist terms, only the rather silly little motive of curiosity as perhaps uniquely human. We can insert this motive as a little wrinkle into the dialectic of survival and say that in science and in the arts, both fine and useful, we get relief from the pressures of natural and cultural selection in amusing ourselves with games, whether in paint or stone, bio-chemistry or mathematics. Yet even science itself is, it is becoming plain by now even to philosophers of science, more than this. Its aim is not only prediction (all that Hobbesism allows it) but *understanding*, that is, responsible self-submission to what, to the best of our powers, we accredit as truth. This side of science, sometimes called 'aesthetic,' already demands ontological levels for its interpretation: much as St. Augustine showed that sense subordinates itself to inner sense, inner sense to reason, and reason to the 'truth of numbers' which it discovers as more authoritative yet than the authority by which it legis-lates for the lower faculties. Now such submission of one part of our nature to another, and to something to which our whole nature in turn submits itself, emerges most clearly, as Miss Murdoch suggests, in relation to our experience of great art. What we discover here, moreover, far from being a little extra on the side, turns out to be a more inclusive actuality which assimilates harmoniously the ultimate meaning of science itself. Thus science as discovery, as submission to what we hope to be the truth, is assimilated to the endeavor of the artist to articulate in sight or sound his vision of the real. The poets are indeed the true teachers of mankind, not only in what they say and in their way of saying it, but in that they embody as synthesizers the fuller range of activities through which man as knower and doer, as seeker after significance as well as maker of significance, finds himself within a meaningful cosmos, a cosmos of which, as maker, he is both subject and sovereign. Not the duality of art

26

against nature, convention against natural force, let alone the sole conquest of the latter, but the identity-in-difference of kingdom *and* cosmos: that is the unity toward which, as we hope, the several questions raised in this volume and the several roads toward their resolution hinted at and partly implemented by our participants convergently aim.

REFERENCES

[1] F. Crick, *Of Molecules and Men*, Seattle: University of Washington Press, 1966, p. 93.
[2] W. Heitler, *Man and Science*, New York: Basic Books, 1963, p. 97.
[3] F. Brandt, *Thomas Hobbes' Mechanical Conception of Nature*, Copenhagen: Levin & Munksgaard, 1928; R. Polin, *Politique et Philosophie chez Thomas Hobbes*, Paris: Presses Universitaires de France, 1953; F. Tönnies, *Thomas Hobbes Leben und Lehre*, Stuttgart: H. Kurtz, 1925; J. W. N. Watkins, *Hobbes's System of Ideas*, London: Hutchinson, 1965.
[4] Watkins, *op. cit.*
[5] R. Cudworth. *The True Intellectual System of the Universe*, London, 1820, IV, pp. 126–7.
[6] T. Hobbes, *Objections III*.
[7] C. G. Hempel, 'On Mathematical Truth,' in H. Feigl and M. Brodbeck, *Readings in Philosophy of Science*, New York: Appleton-Century-Crofts, 1953, pp. 148–62, p. 160.
[8] T. Hobbes, *Leviathan*, Pt. 1, Chap. IV.
[9] P. Tillich, *The Courage to Be*, New Haven: Yale, 1952, pp. 129–30.
[10] Brandt, *op. cit.*, p. 230.
[11] *Leviathan*, Introduction.
[12] C. G. Hempel and P. Oppenheim, 'Studies in Logic of Explanation,' *Philosophy of Science*, 15 (1948), pp. 136–75, p. 136.
[13] C. G. Hempel, 'The Theoretician's Dilemma: A Study in the Logic of Theory Construction,' *Minnesota Studies in the Philosophy of Science*, 2, Minneapolis: University of Minnesota Press, 1938, pp. 37–98.
[14] T. Hobbes, *Elements of Law*, Pt. 1, Chap. II; cf., e.g., Chap. VI and *passim*.
[15] Galileo Galilei, *Discoveries and Opinions of Galileo* (S. Drake, trl. and ed.), Garden City: Doubleday, 1957, p. 274; Galileo Galilei, *Il Sagiattore*, *Opere, Ed. Naz. 6*, 1965, pp. 347–8.
[16] J. L. Stocks, *Reason and Intuition and Other Essays*, London: Oxford, 1939, pp. 62–3.
[17] Watkins, *op. cit.*, p. 79.
[18] Cf. M. Polanyi, *Personal Knowledge*, London: Routledge & Kegan Paul, 1958, p. 390.
[19] J. Bardeen, 'Development of Concepts in Superconductivity,' *Science Today* (Jan. 1963), pp. 19–28.
[20] Cf. G. H. Chew, 'Crisis for the Elementary-Particle Concept,'

submitted to *Science and Humanity* yearbook (Moscow), University of California Radiation Laboratory, Preprint 17137.

[21] S. Koch, below, p. 132.
[22] *Leviathan*, Pt. 1, XI.
[23] C. Taylor, *The Explanation of Behaviour*, Routledge & Kegan Paul, 1964, pp. 120–5.
[24] Polanyi, *op. cit.*, p. 265.
[25] *Ibid.*, pp. 320–1.

PART I

FOUNDATIONS PROBLEMS IN THE SCIENCES

1

EPISTEMOLOGY OF QUANTUM MECHANICS

ITS APPRAISAL AND DEMANDS

Eugene P. Wigner

I believe that in a conference such as ours every participant should contribute some of his specialized knowledge which has a bearing on the main subject of the conference. However, he also should give his thoughts on the main subject even if this is outside his specialized competence, and he should do this freely, not restricting himself to the areas of impact of his speciality. This may give a dilettantish taste to some of his remarks, and I wish to apologize in advance for my own. The specialty which I wish to contribute lies in the area of the epistemology suggested by our present picture of the conceptual limitations of physical theory. However, I would like to come to this subject, and its place in our subject of inquiry, from a more general discussion of cultural unity and the role of science in general therein.

THE EXPANDED ROLE AND GOAL OF SCIENCE

Forty years is not a very short period in human affairs, and it is likely that some aspect of man's world has undergone important changes in almost any period of forty years. Our attention is, naturally, focused on those aspects of our world which are changing, and I believe that the changes in the role and also in the goals of science are as characteristic of our times as any.

Forty years ago most people knew about science only as something esoteric. When I was a child there was only one scientist in the circle of acquaintances of my family, and he was considered to be somewhat queer. Today in the U.S. one person in thirty works directly or indirectly either on improving our understanding of nature or to make better use of the understanding which we possess. Forty years ago, the demands of

31

science on our economy were negligible—the most spectacular ones being for astronomical equipment. Today the U.S. spends $20 billion a year on research and development. This is $3\frac{1}{2}$ per cent of the gross national product.

Forty years ago few people paid much attention to what scientists thought or said; today their voice carries great weight in national as well as international affairs—often uncomfortably great weight. The attention, also, which the world pays to scientific discoveries or observations has grown to an almost disagreeable extent, and a significant scientific error could—in fact, occasionally does—cause embarrassment to its country of origin.

Hand in hand with this expanding role of science went an expansion of its goals. Forty years ago science was happy to provide increased insights into very limited areas—the motion of celestial bodies being a prime example. Today science seems to strive for an encompassing view of the whole universe in all of its manifestations, both in the large and in the small.

As I hope to explain on some other occasion, the expanded role and the expansion of the goals of science hold great promises for the future of man but also bring a new and more subtle type of danger thereto.[1] The promise is, of course, that men, instead of fighting with each other for power and influence, will fight together for an increase in our knowledge and understanding. If there is to be a cultural unity among men, science will have to provide most of the pillars thereto.

THE PRESENT SCHISM IN SCIENCE

What is the most important gap in present science? Evidently, the separation of the physical sciences from the sciences of the mind. There is virtually nothing in common between a physicist and a psychologist—except perhaps that the physicist has furnished some tools for the study of the more superficial aspects of psychology, and the psychologist has warned the physicist to be alert lest his hidden desires influence his thinking and findings.

Yet psychological schools have maintained that they wish to explain, eventually, all 'processes of the mind' by known laws of physics and chemistry and, as I'll enlarge upon later, physicists came to conclude that, in ultimate analysis, the laws of physics give only probability connections between the outcomes of subsequent observations or content of consciousness. Hence, there is a striving, on the part of both the psychological and the physical sciences, to consider the reality of the subject of the other one the more basic. Perhaps I should interject here that it is my belief that the endeavor to understand the functioning of the mind in terms of the laws of physics is doomed to have no more than temporary and very partial success, whereas the direction into which modern physical theories point appears to me to be more fertile. Fundamentally, however, I believe that physics and chemistry, the disciplines of inanimate matter, will prove a limiting case of something more general, and that the fault of our ancestors in having thought of body and mind as separate was not of having thought of both of them but of having thought of them as separate. They form a unit, and both of them will be understood better if they are considered jointly.

This is, of course, only an opinion and a belief which may not be fully appropriate to our subject. What I feel sure of, however, is that there is a vast area of interesting knowledge here, waiting for a great mind to start its uncovering. To have such a large area may be very important if science is to become a unifying force for men, diverting them from their preoccupation with power.

TWO TYPES OF SCIENCE

The hope that man can fill the gap between the physical sciences and those of the mind is inspiring. It can be put into proper perspective, however, only if we also discuss what 'filling the gap' means, i.e. what it is that science accomplishes when it pervades a subject. Before that, a distinction between two kinds of science should be drawn which is surely neither new nor precise but which is nevertheless useful. Some of our sciences, among which physics is foremost, are concerned solely with regularities. The typical statement it makes is something as follows: If two macroscopic bodies are far from each other and all other bodies, the component of their separation in any fixed direction is a linear function of time. This is, evidently, a rule of very great generality (even though it is a special case of even more general

33

'laws'). It is, however, a highly conditional statement and leaves unanswered many questions, such as the character of the bodies which are widely separated, whether there are such bodies and how many such bodies exist. There are other sciences which are concerned with just these questions, not with regularities. Geography, astronomy, botanics, zoology, and at present also psychology are such sciences. They are descriptive rather than searching for regularities, even though, as they develop, they may discover regularities—as did astronomy. However, these regularities then seem to become parts of another discipline.

LIMITATIONS OF THE TWO TYPES

Several years ago I discussed the limits of the sciences concerned with regularities. My conclusion was that the limits are given by the finite capacity of the human mind for assimilating knowledge, by its finite interest for increasingly subtle and sophisticated theories. Some of these limitations can be overcome by co-operative science, and there is a great deal of temptation to discuss that. The limitations of the descriptive sciences have evidently the same source. Even if we could catalogue the exact locations of all houses and trees all over the earth, we would have little interest in such a catalogue. It surely would be interesting to know how the plants look on another planet on which the conditions, such as temperature, gravitational acceleration, etc., are different from ours; it would be even more interesting to know how they look on a planet which is very similar to ours. But, unless comparisons of plants on different planets suggest new regularities, interest in such comparisons would soon be exhausted. It seems to me, therefore, that if the sciences are to provide the pivot for cultural unity, we must rely in the long run principally on the sciences which discover regularities, or, as is said commonly, provide explanations.

Even the regularity-seeking sciences may become stale one day; even they may cease one day to fire man's imagination. The ideals of Arthur's Round Table did. Man may then turn to other ideals. However, the fascination of the regularity-seeking sciences should suffice to give man a taste of cultural unity; it should start him on that path.

EUGENE P. WIGNER

WHAT DO THE REGULARITY-SEEKING SCIENCES FURNISH?

It is often said that physics explains the behavior of inanimate objects. This sounds somewhat like an advertisement; no ultimate explanation can be given for anything on rational grounds. As I have said elsewhere, what we call scientific explanation of a phenomenon is an exploration of the circumstances, properties, and conditions thereof, its co-ordination into a larger group of similar phenomena, and, above all, the ensuing discovery of a more encompassing point of view. Or, as David Bohm said a short time ago, 'Science may be regarded as a means of establishing new kinds of contacts with the world, in new domains, on new levels. . . .'[2] One of the greatest accomplishments of physics in this regard is also one of its oldest accomplishments: the recognition that the motion of an object thrown into the air, the motion of the moon around the earth, and the motion of the planets around the sun follow the same regularities. The paths are all ellipses, obeying Kepler's laws. The more encompassing point of view was provided in this case by the gravitational law of Newton, which permitted the co-ordination not only of the three phenomena enumerated but also of several other less striking phenomena.

The regularity-seeking sciences have another function: the discovery and creation of new phenomena. The phenomenon of electric induction is a case in point. Its co-ordination with magnetic phenomena, with the forces exerted by currents, was possible under the point of view of the Faraday-Maxwell electrodynamics.

These examples are brought forward to indicate what 'filling the gap' between the physical sciences and the sciences of the mind should mean. We have at present, of course, a rather sophisticated and well-developed regularity-seeking science of inanimate objects. We also have the beginnings of a similar science of the mind. Surely, the concept of the subconscious has permitted us to see many phenomena from a common point of view and the theory of the subconscious has been much enriched by phenomena discovered more recently. Dr. Polanyi has alluded to these. However, the link between the two, the phenomena of the mind and of physic phenomena, is missing now, just as the link between gravitation and mechanics on the one

35

hand and electromagnetism on the other was missing for a long time. The creation of the missing link is a sufficiently challenging task to form a pivot for the cultural unity which we dream about.

Before turning to the subject about which I am believed to know something, the epistemology of modern physics, I would like to make one more general remark supporting a point made by Dr. Polanyi.[3] It has little to do with cultural unity but much with the epistemology which I want to discuss.

SCIENCE IS AN EXTENSION OF PRIMITIVE KNOWLEDGE; IT IS IMPOSSIBLE WITHOUT THE LATTER

In order to appreciate this point we only have to imagine a mind which knows all that we know of physics but knows nothing else. Such a mind is like that of a person who floats in dark empty space and is not subject to any outside influence. Such a mind will have no use for its abstract knowledge of physics because there will be no sensory data to be correlated, no events to be understood. In fact, the laws of physics will be meaningless for him, because these give correlations between sensory data and he receives no sensory data that he can interpret.

This is very abstract discussion, but it does show that science cannot be a replacement for our common ability to accept sensory data, an ability mostly born with us but also partly learned during our babyhood. Rather, science gives us only a different view of these sensory data; it creates pictures from which they can be correlated in novel fashions. The primitive sensory data are the material with which science deals, which it orders and illuminates. Science does not furnish data, neither does it offer a substitute for the data; it is only interested in them.

One could object that our knowledge of nature permits us to use substitutes for sensory data, to photograph the stars rather than to look at them. This, however, is only appearance. Even if we photograph the stars, we must eventually 'take in' by our senses what the photograph shows. Furthermore, without our senses we could not handle a photographic camera. Clearly, all knowledge comes to us ultimately through our senses; science only correlates this knowledge.

I make these remarks for two reasons—neither of which is directly related to 'cultural unity.' First, because I believe that they are a paraphrase of Dr. Polanyi's insistence on the significance of tacit knowledge; in fact, they perhaps are somewhat more. They are an insistence that something even more primitive than tacit knowledge is the subject of science, the material it deals with, without which it would be empty and meaningless.

The second reason for my remark is that the same conclusion at which we have arrived here abstractly will be forced on us later when we consider the epistemology of modern physics. This is not surprising; in fact, obstacles which can be understood abstractly only with foresight and imagination become obvious if one tries to travel down the road on which they stand.

THE TREND IN PHYSICS IN OUR CENTURY;
THE CONCEPTUAL FRAMEWORK OF QUANTUM THEORY

Physics in our century has been under the spell of two conceptual innovations and an experimental discovery. The two conceptual innovations are relativity and quantum theories; the experimental discovery is the realization that the structure of matter is atomic, that it can be explored with tools developed for this purpose; pictures of its constitution, enlarged to macroscopic scale, can be obtained rather directly.

Let us first look at an element that is common to relativity theory, as it was originated by Einstein, and to quantum theory as it developed, often against the better wishes of its disciples. The common element—an element which is very important from the epistemological point of view—is the rejection of certain concepts which have no primitive observational basis. These concepts are very different from the two theories, and I prefer to discuss quantum theory principally because its rejection of concepts not based on direct apperception is much more radical. Let me therefore describe present quantum theory's critique of the earlier, very natural, concepts.

The quantities which characterize the state of a system of point particles were considered, in analogy to macroscopic objects, the positions and velocities of these particles. An analysis of the experimental processes available for determining atomic structures—the processes which were mentioned before as constituting the most important experimental discovery of our century—shows, however, that there is little reason to believe that these can be determined accurately on the atomic scale. Hence, quantum theory wishes to adopt an attitude which is free of preconceptions. It considers the observations themselves.

An observation implies the interaction of some 'measuring apparatus' with the 'object' on which the observation is to be undertaken. One can think of such an interaction as a collision between the apparatus and the object. Just as in the case of a collision, there is practically no interaction before or after the process; before and after the observation, apparatus and object are isolated from each other. The duration of the interaction is finite; it is often idealized as infinitely short. It results in a statistical correlation between the states of apparatus and object.

Since the concept of observation (also called 'measurement') plays a basic role in the epistemology of physics, it may be useful to furnish an example of the kind of interaction that we have in mind. The example to be given assumes the validity of classical mechanics which will be more familiar to most people than its quantum counterpart. Since the purpose is only to illustrate the role and general character of observations or measurements, this should not matter. Although it is true that the analysis of the role of observations is not necessary in classical theory, such an analysis remains valid, and it is hoped that the absence of quantum effects will render the discussion more visualizable.

As 'object' let us consider a ball which can move in a horizontal trough; this is our object. We do not know where it is or what velocity it has. In order to measure or observe its position we may use, as apparatus, another ball which is moving in the same trough. This we can roll at a known speed toward our object-ball. It will collide with it, return to us, and if we obtain its speed again and the time of its return we can calculate when and where it collided with our object-ball. Even before it returned to us, its position depended on the position of the object-ball—there was a statistical correlation between the two.

It would have been, perhaps, a bit more appropriate to use a light or radar signal for measuring the position of our ball in the trough. Light or radar signals are not so close to direct experience, but they do have the advantage that their speed need not be determined—it is light velocity. Hence, only the time of return needs to be ascertained in order to obtain the position of the ball at the time just midway between firing and receiving the light or radar signal.

This is, of course, a very primitive example. It does show, however, all the essential elements of an observation as conceived in present quantum theory. One needs an apparatus—the second ball or the radar signal emitter—and one has to read the apparatus in the same sense as one reads a voltmeter or ammeter. In the case considered, one has to ascertain the time of return of the ball or radar pulse. The example also shows all the weaknesses of the theory. The first one is connected with the 'reading' of the apparatus, that is, ascertaining the time of return. If I could not directly ascertain the position of the object-ball, why should I be able to ascertain the return time of a signal? How about the apparatus? How did I send out a signal at a sharply defined time?

These are two difficult questions which relate very closely to the theme which I discussed under the heading 'Science is an Extension of Primitive Knowledge.' Before turning to this connection, I would like to describe the language in which the laws of physics should be formulated—and are actually formulated in quantum theory—if we adopt as general a view on the nature of observation as indicated above.

THE LANGUAGE FOR THE LAWS OF NATURE

We shall now seek to determine what kind of use can be made of the results of observations, how they permit us to foresee the future, at least partially. Let us turn to the example of the determination of the position of the ball in the trough. The answer is that having any three such determinations, finding the positions x_1, x_2, x_3, with the corresponding times t_1, t_2, t_3, there is a linear relation between these so that

$$x_1 = vt_1 + a \qquad x_2 = vt_2 + a \qquad x_3 = vt_3 + a$$

is valid for all three pairs x,t with the same v and a. These equations simply express Newton's law for the ball in the trough: that it persists in its uniform motion. This assumes, of course, that the reflection of the radar signal did not influence the motion of the ball—an assumption which is possible in classical theory, which deals with macroscopic objects, but would not be permissible in quantum theory, which deals with atomic particles, the motions of which are significantly influenced by the reflection of a signal.

The fact that all three equations between x,t pairs are valid with the same v,a can be expressed also by the statement that the determinant

$$\begin{vmatrix} 1 & 1 & 1 \\ x_1 & x_2 & x_3 \\ t_1 & t_2 & t_3 \end{vmatrix} = x_2 t_3 - x_3 t_2 + x_3 t_1 - x_1 t_3 + x_1 t_2 - x_2 t_1 = 0$$

vanishes. This is then the formulation of the law of nature (in this case of Newton's first law) which contains only observed quantities; it gives a correlation between the results of several observations. One can, perhaps, even further emphasize the observational character of the quantities appearing in the last equation by writing for each x the difference between the time of emission t_e and time of return t_r of the radar signal, divided by c. Similarly, each time t can be replaced by the average $(t_r + t_e)/2$ of the time of emission and return. This gives after some calculation:

$$t_{e2} t_{r3} - t_{e3} t_{r2} + t_{e3} t_{r1} - t_{e1} t_{r3} + t_{e1} t_{r2} - t_{e2} t_{r1} = 0,$$

where t_{e1} is the time of the emission of the first signal, t_{r1} the time it returned, and so on. The last equation contains only quantities which have a simple observational significance, at least as long as we suppose that the times of the emission and receipt of radar signals can be directly measured.

Whereas the formulation of Newton's first law, given in the last equation, may not be the most natural one, the laws of quantum physics can very naturally be formulated in terms of observations. This seems to me to be a distinct epistemological advance; after all, a law of nature is, as its name indicates, an expression of correlations, of correlations between observations or results of observations. The observations considered in quantum theory are of much—in fact, infinitely—greater variety than the single, rather primitive type of observation which we considered: the emission and return of a radar signal and the measurement of the corresponding time.

If we accept the principles of quantum mechanics, it can be shown that each observation can be characterized by a self-adjoint operator in a suitable Hilbert space—the characterization von Neumann gave to his 'physical quantities.' It can also be shown that the inference which past observations permit one to draw about the outcomes of future observations can be fully characterized by a vector in said Hilbert space, that is, a 'wave function.' However, it is dangerous to attribute physical reality to this vector, first, because it is not quite clear what physical reality means, and second, because it changes as a result of observations in a way not given by its equations of motion. This is called, in the jargon of quantum theory, the collapse of the wave function. In fact, as a result of the possibility of statistical correlations between isolated systems (such as the correlation between measuring apparatus and object), components of the wave function which refer to one system can change as a result of observations carried out on another system.

The remarks of the last paragraph were meant to establish the connection with other terminologies of the literature for those who are familiar with that literature. We now proceed to a critique of the epistemology just outlined.

THE PHYSICISTS' CRITIQUE OF THE CONCEPTUAL FRAMEWORK OF QUANTUM PHYSICS

The observation described in the preceding section has two elements outside the observer—the apparatus, which was the radar emitter in our example, and the 'reading' of this apparatus. This was the measurement of the time of return of the radar signal.

Clearly, neither of these elements is outside the scope of physics. In fact, apparatus for the physicist is made according to prescriptions furnished by the physicist, if not by the physicist himself. Similarly, we teach the physics student how to read the apparatus. Neither of these two processes can be considered as truly primitive, unanalyzable, in a realistic theory.

The answer to the preceding objections is that it is possible to reduce both processes, that of the use of the apparatus and that of its reading, to simpler ones. Let us consider the reading first. This is, essentially, an observation of the state of the apparatus after its interaction with the object. Such an observation

41

can be carried out in a way similar to the original observation by considering the apparatus used for the original observation as the *object* to be observed by a second apparatus. This second apparatus can be made to enter into a temporary interaction with the first apparatus which results in a statistical correlation between their states. Instead of reading the first apparatus, the time of return of the signal in our case, one can read the second apparatus, perhaps the position of a dark spot on a photographic plate. If this is not yet considered as a primitive observation, one can go further and consider the eye of the observer as a third apparatus with which the second one is 'read,' i.e. its state observed. This is a consistent scheme as far as it goes: the statistical correlations between object and first apparatus, and between first and second apparatus, entail a statistical correlation between object and second apparatus so that reading the latter is equivalent to reading the first. This is very satisfactory. It should not obscure the fact that, nevertheless, an ultimate reading is unavoidable: there must be something that is *directly apparent* to the observer, though it is not necessary to specify what this is. It may be the state of his retina, it may be the position of the spot on the photographic plate, and so on.

This transferability of the dividing line between observer and apparatus was recognized by von Neumann. It has attractive as well as perturbing aspects. It is satisfactory that the whole scheme is consistent, that the expression given by quantum theory for the probabilities of the outcomes of observations guarantees that the same results are obtained no matter what method, or how direct or indirect a method, one uses for the observation. It is perturbing that there is no definite limit to which one can and should follow the transference of the information of the state of the object. Toward the end, when looking, for instance, at the photographic plate, one makes use of the tacit knowledge of Dr. Polanyi. This could be seen to begin with as being ultimately unavoidable. It is nevertheless an element which is foreign to the otherwise precise and clearly articulated framework of the theory.

The situation with the apparatus is similar but perhaps even more striking. Naturally, we cannot use an apparatus unless we know its properties, whether it is a voltmeter or an ammeter.

42

This knowledge is not born with us. To many of us it does not even come very naturally. The orthodox answer of the quantum theorist is that we should subject the apparatuses to observations which will tell us their properties. This again leads us to a chain, because in order to observe the properties of one apparatus, one must use other apparatuses, the properties of which are also unknown to begin with. In fact, the chain is not a simple one in this case but will have increasingly many branches. And in fact this is not the way one proceeds—one trusts a colleague who tells us what equipment to use and whether it is in good order. This method of ascertaining the properties of the apparatus is, however, clearly outside the framework of physical theory and illustrates, even more strikingly than the situation encountered with the reading of the apparatus, that present physical theory is not self-contained, that it constantly relies on everyday knowledge and is only an extension thereof.

The physicist has a few more, rather technical, points of concern with the details of the theory of observation. I mention this only in order to avoid the impression that I consider those difficulties to be solved. These difficulties concern the assumption of the instantaneous nature of the measurements and the determination of the class of measurable quantities. In my opinion, these difficulties are of no concern to us now.

The present section dealt with the difficulties which one experiences as a physicist with the conceptual framework of present physical theory. Briefly, the boundaries of physics become indeterminate, and the extent of reliance on common experience is unclear. The early days of lack of concern with the conceptual foundations are gone. This may give us some nostalgia for bygone days, as does all change and progress. However, progress is also refreshing, and the evident need to be concerned with the foundations of physics, which lie partly outside of physics, may be a harbinger of more intimate connections with other sciences.

THE NON-PHYSICIST'S CRITIQUE OF THE CONCEPTUAL FRAMEWORK OF QUANTUM PHYSICS

The laws of nature are formulated by the quantum physicist as probability connections between outcomes of observations.

Some of these observations may be simultaneous; others will take place in succession. There is a relatively simple formula for the probability that observation Q_1 yield the value q_1, that Q_2 yield q_2, . . . and so on. This probability is expressed in terms of q_1, q_2, . . . and the self-adjoint operators of Hilbert space which correspond to the measurements Q_1, Q_2, . . . and the times t_1, t_2, . . . at which they are undertaken. Hence, all that remains to perfect is the coordination of proper self-adjoint operators to suitably defined measurements!

The difficulty which the non-physicist and the physicist alike find with the preceding picture is that it 'idealizes' the observer, who is a man of flesh and blood, as a data-taking automaton, never forgetting anything, having no other knowledge but that obtained from his observations.

Obviously, this is too schematic a picture. Even as thinking men, we usually think in terms of everyday concepts—in fact, mostly tacitly, as Dr. Polanyi has explained. Without any mathematical support, there is an amazing consistency in the simple everyday observations. As I remarked once, if our keys are not in the pocket where they should be, we do not suspect a new phenomenon, but go to look for them where we may have forgotten them. Furthermore, we stop looking for them on the second floor if our wife tells us that she found them on the first. Clearly, most correlations between observations are not obtained by the formulae of quantum physics.

Even more important, it seems to me, is the fact that we are not only cognitive beings, that correlations between cognitive functions give only a very small part of the life of our minds. The other parts will interfere with the cognitive functions so that these cannot be considered independently and in separation from the others. This is where I return to the general discussion of the schism between the physical sciences and those of the mind for which I had expressed the hope that it would become less deep than it now is.

One of the most important past accomplishments of physics is that it specified the realm of the explainable. This comprises the second and higher time derivatives of the positions with respect to time, not the values and first derivatives. No abstract insight could have given this information. Similarly, the realm of the explainable in the generalized science will have to be discovered—we are far from knowing it now. The knowledge of this realm would tell us in which areas we can hope to find new and interesting illuminations. Finally, we will have to draw some limits of how much we want to know, how much we consider to be interesting. These are formidable tasks, and we can only hope that the human mind is equal to them.

REFERENCES

[1] See *Symmetries and Reflections* (Indiana University Press, Bloomington and London, 1967), article 24: 'The Growth of Science—Its Promise and its Dangers.'

[2] D. Bohm, *Special Theory of Relativity*, New York: Benjamin, 1965, p. 230.

[3] M. Polanyi, 'The Creative Imagination,' to be published in *Psychological Issues*, 1967.

GOETHE AND THE PHENOMENON OF COLOR

Torger Holtsmark

Goethe's extensive color studies, assembled in his *Farbenlehre* (1810), and consisting of three parts: historical, didactical, and polemical, have long been obscured by the successful development in the natural sciences, as first advanced by Newton's 'Opticks' (1704).[1] Goethe's vehement polemics against Newton's 'Opticks' helped to discredit him in the eyes of the scientists. Still a certain number have felt something inexplicable in the circumstance that a man of Goethe's intellectual dimensions should take a standpoint opposed to the leading school of progressive physics. Various attempts have been made to explain the divergent viewpoints of Goethe and Newton.[2]

One of the most typical, as well as the most uncompromising, is that of Helmholtz in his classical treatise on physiological optic:

His [Goethe's] concern is not to develop a physical explanation of the color phenomena. As such his postulates would make no sense. What Goethe tries to do is to demonstrate the general conditions for color appearance. These conditions he expresses in terms of a primary phenomenon. . . . His vehement polemics against Newton seem rather to have their source in that the hypothesis of the latter appears absurd to him, than that he has anything against Newton's experiments and the conclusions drawn therefrom. But the reason for Newton's assumption that a white light is composed of many colored lights appearing so absurd to him is to be found in his particular point of view—that of an artist—which leads him to seek beauty and truth directly expressed in concrete images. At that time the physiology of the eye had not yet been studied. The composite nature of white asserted by Newton was the first decisive empirical step toward seeing the subjective content of sense experience, and Goethe was justified in his premonition when he so vigorously opposed it,— it seemed to him to destroy the 'fair illusion' of sense experience ('Der schöne Schein').[3]

Anyone who has discussed these questions with practicing physicists may recognize the arguments. They arise from the general feeling that Goethe presents us with a world vision that does not satisfy the scientific demand for objectivity and exactness. But we may also recognize a certain oversimplification of the problem involved. Helmholtz dismisses Goethe's *Urphänomen* without really explaining its nature. He refuses to treat Goethe's world vision as a matter for serious consideration from the physicist's point of view, even though he accepts Goethe's observations. He touches upon the relation between the artist and the scientist without considering that Goethe himself was seriously occupied with that relation and mentioned it more than once.

Helmholtz refers to a certain *experimentum crucis* in support of Newton's hypothesis, an experiment of which Goethe supposedly had no knowledge, but he does not mention what experiment it really was or how it might be regarded from Goethe's point of view. Since Helmholtzian days there has been much discussion about the role of crucial experiments within the progress of physics, and it has been demonstrated that in our appraisal of an experiment a certain style of thinking, a certain terminology operates, influencing what we suppose we observe. Goethe himself was the first to admit that this state of things applies to the disparity between the Goethean and the Newtonian approach. Thus in the introduction to the polemical part of his color theory he states that his aim cannot be to 'prove' the failure of Newton's treatment, but to demonstrate that its singular style of thinking does not do justice to the phenomena themselves.

Helmholtz sums up his objections by postulating that Goethe's aim is to reestablish the 'fair illusion.' Listening to Goethe's own words, however, this is not the impression we get. Goethe gave the most serious consideration to the nature of scientific activity, and for long periods he accorded this occupation the highest priority. He regarded his color theory as his most important work, compared to which he said that even his poetry must take second place. The primary motivation for his scientific studies were part and parcel of his search for human self-education. Hence his efforts to extend his color theory in the form of practical material for the sake of self-study:

My color theory, as you remember, should not be read or studied only. It should be done.[4]

Goethe's color theory is a complete work in itself, hewed out of concrete experiences and concise phrases. It can only be discussed after a critical examination of those experiments and phrases. Only then should we permit ourselves to pose the question which Helmholtz could afford to have asked himself, namely: What can I discover in Goethe that I cannot find in Newton?

In a letter to Schiller, Goethe spoke about his own scientific activity:

My considerations of Nature give me much pleasure. It may seem curious and yet natural that out of them arises some sort of a subjective whole. It might be said that it is the world of the eye which is exposed through color and form. If I examine my own actions, I recognize that I use primarily the organ of sight, whereas the other sense organs are only rarely brought into action. All reasoning gives place to description.[5]

The expression 'world of the eye' is a clue to the Goethean view of the world. Goethe's profound interest in the activity of his senses determined his extensive and systematic work in the various fields of natural history together with his approach to experimental methods. To see the thing himself, to be able to talk about things as something he had seen and experienced, became a fundamental motive in Goethe's personal development. He was convinced that this must be bound up with a specific human aspect of his existence. His scientific activity was motivated by the idea that the individual participates in the world through the actions of his senses, since man establishes himself as an individual being through the way he deals with his own experiences. The human world emerges through the act of experience; it is therefore something that becomes and which is never finished. The world that *has* become is already behind the individual, and only in so far as one considers the world as something that has become, could one be tempted to characterize it as something that belongs to oneself, as when I use expressions like 'my world,' for instance. The experienced world cannot be 'my' world, since it has not yet become. It is even 'the world of the eye.' Thus, in Goethe's view, the human world is not a subjective one. An eventual subject in that world would have to be experienced like all the other things I know about. The act of experience, therefore, is situated over and above the limits of the subject.

It is not easy to express the divergence between the Newtonian and the Goethean attitude. Obviously both of them are looking for the dynamics of color phenomena; they will bring color back to the primary principles for its appearance. But we may say that Newton is inclined to consider color as something that has become. His colors correspond with a limited set of well-defined physical parameters. In Goethe's view color is something undetermined and undeterminable, something that arises through the interplay of certain dynamic conditions and is moved in one direction or other according to the changing balance of the creative conditions. To see a color as a result of that dynamic interplay would be to see it within a *pure phenomenon*, an *Urphänomen*.

To rest at last in front of the pure phenomenon is admittedly an act of resignation. But still it is the difference between resignation before the limits of mankind, and a resignation inside the hypothetical limits of my narrow-minded individuality.[6]

Goethe felt compelled to give his thoughts on scientific activity as a pragmatic approach to the pure phenomenon a more concise form when he wrote to Schiller, who had more philosophical training than Goethe had. His basic propositions were more or less identical with what has been recognized and practiced as the empirical and experimental method since the Renaissance. This is the well-known shuttle movement between the special and the general, between the single phenomenon and the law. What gives the Goethean treatment its specific character and flavor does not arise from the formal procedure; it is, rather, inherent in his attitude toward experience as such. He writes:

As a result of our work we should be able to demonstrate:

1: *The empirical phenomenon:* That which all men can observe in Nature, and which in its turn is sublimated into

2: *The scientific phenomenon* through experiments, described under viewpoints and in contexts and conditions different from those

50

under which it was first observed, and exhibited in a more or less successful progression.

3: *The pure phenomenon* emerges as the first result of all experiences and experiments. It can never exist isolated in itself, it reveals itself in a never-ending succession of observable facts. In order to describe it, the human mind determines what is empirically imprecise, sorts out what is sporadic, separates the impure or composite, develops that which is in a state of turbulence and even explores the unknown.[7]

In a gently self-ironical way, Goethe relates how at one point of his life he was compelled to study the laws of painting in the hope of thus arriving at a more conscious understanding of the art of poetry, which he had hitherto mastered unconsciously. He therefore questioned his painter friends about the criteria on which their art was founded, but he soon discovered that they could no more express this than he himself could formulate the rules of poetry. Thrown back, therefore, on his own resources, he conceived the idea that the laws of using color must be sought in the objective phenomena of natural colors themselves. From that time on he devoted time and attention to the color processes in nature. In this connection he relates an anecdote that throws light on the consistency of his thought. In the course of his concentrated study of the appearance and disappearance of colors in nature, his attention had been drawn to a certain relationship between darkness and the color blue. Probably he was thinking of the gradual changing of the black color of the night to the blue of dawn or day. Now since it may be rightly said that darkness is not a color, Goethe declared, half in earnest (in order to stir up his friends a little): 'Blue is not a color.' This was symptomatic, since it shows how Goethe from the beginning directed his attention toward the inner relations between colors, and also that he reflected on the phenomenon of darkness in its relation to the appearance of color.

It is also interesting that while Newton's theory about the composite nature of white light was well known to Goethe, this did not interest him very much. It did not appear to him that the solution of *his* problem could be found within that theory. That Goethe would have little in common with Newton's doctrine of atomism is easy to understand. But why was not the disagreement between them already obvious at that time? Subsequent developments show us that an important factor in Goethe's attitude to Newton was still lacking.

At this point the well-known incident of Goethe borrowing prisms, etc., from Hofrath Büttner took place. Goethe wanted to make the Newtonian experiment for himself. But the prisms remained untouched, perhaps understandably: if one is primarily concerned in the color of things, there seems little point in looking at them through prisms, which produce a distorted view; better to rely on one's own healthy eyesight. Thus the dust gathered by the prism cases under the bed reflects Goethe's lukewarm attitude toward experimental physics. However, Hofrath Büttner, an orderly man, called for his prisms, and at the last moment Goethe took a quick look through them. This was to have far-reaching consequences in his life. In one moment Goethe was born as an experimental physicist, and there was no more talk of returning the apparatus. He saw something curious. What was it?

We shall not go into detail at this point. But two things have to be considered. Goethe's first idea was to look at the white wall through the glass prisms, as he supposed that the wall would appear in the seven colors of Newton's spectrum. This was due to a certain misunderstanding. According to Newton's explanation, a 'spectrum' will appear when only a narrow beam of light is entering the prism, a condition which is not fulfilled when an extended surface is looked at, as Goethe was doing. Goethe was surprised as the wall seen through the prism remained white.

Now Goethe turned round and looked at the window. Seen with the naked eye, the frame and the cross-piece appeared as a black sign against the whitish-greyish sky; but when he looked through the prism, lively colored transitions appeared along certain border lines between light and darkness. This appearance was unexpected, and Goethe felt that here was part of the answer to the principal question that had drawn him into the serious study of color, namely the question about the objective laws of coloring. The colored transitions he was now observing were not part of the coloring of things as such. This was seemingly a 'free' color phenomenon, but it showed the way colors group themselves together according to some law. It seemed to Goethe that by a lucky accident he had touched upon the *Urphänomen* of color appearance.

This episode may have influenced Goethe's ideas on the nature of scientific discovery. He had posed a question which he was inclined to connect with the activity of the eye itself. Now he got part of the answer from an objective physical experiment where the instrument, in this case the prism, operated according to the same optical laws as those operating in the eye. The prism functioned as an elongation of the eye itself.

The didactical part of Goethe's *Farbenlehre* is divided into six main chapters:

1. 'The Physiological Colors'
2. 'The Physical Colors'
3. 'The Chemical Colors'
4. 'General Ideas on the Nature of Color'
5. 'Relations to Other Fields'
6. 'The Moral and Aesthetic Action of Color'

The first chapters deal systematically with color appearance from three fundamental points of view. The final chapters deal with color both from a more pre-scientific and a more post-scientific point of view. The didactical aim of Goethe's color theory must be supposed to be reflected in its main composition, and in fact we know from Goethe's correspondence with Schiller how seriously Goethe considered the question of the composition. His actual composition differs from the general practice in color theory since Newton's 'Opticks,' which is to begin with a hypothetical definition of the physical nature of light and color. Feeling that such definition would mean elimination of color as something emerging, color as a pure experience, Goethe did not do this.

A phenomenon fluctuates. It comes and goes. *How* does it come and go? The answer can only be descriptive, because phenomena are for ever changing into other phenomena. This transformation is their coming and going. The transformation is the phenomenon of the phenomenon, or the 'pure' phenomenon. In a certain sense we may state that Goethe's *Urphänomen* is the dynamic principle of the phenomena. Thus in the systematic part of his color theory Goethe deals with the dynamics of color phenomena from three fundamental points of view. These three active principles are the eye, the light and the substance. Normally color does not appear determined by one principle only. Color is mostly the color of something, that gets its color according to the state of light, and is seen by a living eye.

The physiological colors. The aim of this chapter is to develop a certain idea concerning the nature of the seeing eye. The word 'eye' in this connection has a wider meaning than the mere anatomical and physiological one. Goethe's idea of the eye is related to ideas developed in his morphological writings, namely, that an organism is built up by a hierarchical set of functions ranging from its particular and peripheral functions, to the higher and central functions which operate through a certain equilibrium between the particular functional processes of the environment and the particular functional processes of the organism itself. Thus through the higher functions the organism states itself as an autonomous being. In this sense the seeing eye also is considered by Goethe as an autonomous organism. This idea shines through when he talks about the eye as an 'organ of opposition.' The eye manifests itself as an autonomous being by setting itself up against environmental influences. Through this act of opposition the eye establishes an inner world within the world.

A striking demonstration of an inner dimension within the visual picture is the fundamental distinction between the achromatic (greyish) hues and the chromatic ones. Whereas the achromatic hues together form a linear multitude ranging from black to white, the structure of the chromatic continuum is a circular one. The chromatic hues group themselves together along a closed curve, the 'color circle.' Such topological structures cannot be derived from the laws of the outer world; they have their origin in the lawful functioning of the eye. Thus Goethe aims to describe the eye through its own actions.

'Physiological' color phenomena are subdivided into two classes: the 'normal' and the 'pathological.' The former belong to the sound eye, constituting the necessary conditions for the process of seeing, which they reveal through their mutual interactions as well as through their interactions with more objective color phenomena. The pathological colors deviate from the normal course, and it is thus that they throw light upon the former.

54

Goethe begins with a study of the reaction of his own eye to light and darkness in general. He exposes himself to strong light, which produces not only the impression of dazzling luminosity, but also a certain breakdown of the visual sensitivity. Moving from this state of things into a darkish room he gets the impression of total darkness. Only gradually does he become aware of the details. Only gradually does the eye become capable of establishing that range of shades which together constitute a 'picture.' The eye adapts itself to the level of lightness, as we say, which according to Goethe's terminology means that a certain equilibrium has been established between the eye and the environment. What the eye experiences just after having left the lighted room is more or less the same impression as when it is in absolute darkness. It is not really the color black, since the genuine black color only arises when lighter shades are also present in the visual area. The impression of total darkness (*eigengrau*) can hardly be classified as any specific quality (sense datum), since it cannot be compared with anything else. Still we may recognize it as a result of the pure intrinsic activity of the eye, since when this impression arises after a prior exposure to strong light, it suggests that the eye has responded to the ultimate point of its intrinsic activity in order to withstand a pressure from without. Generally, in so far as a visual picture consists of a wide range of lighter and darker shades, it is formed through a certain equilibrium between activity from without and activity from within. The darker shades of the picture may be considered as the traces from the inner activity of the eye. Hering has made an interesting demonstration of this. If I look at a printed page in a dark corner of the room, the letters are difficult to recognize as all the shades are more or less greyish. But if I go to the window, the letters take on a deep black color although they are now reflecting much more light. Thus between the darker and the lighter shades of the visual picture there exists a dynamic relationship that arises from the efforts of the eye to arrive at a state of balance with the external world. The ability to see breaks down at very high as well as at very low lightness levels because at these levels the state of balance is disturbed. The eye can only fulfill its purpose at lightness levels where a wide range of fluctuating equilibrium can be established. Under such conditions the momentary appearance of one and the same surface will change according to the varying conditions, as for example in the following arrangement:

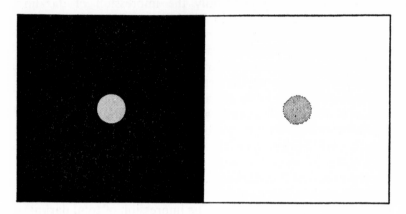

Although the central areas are the same, they appear with different hues. Their apparent darkness depends on the lightness or darkness of the surrounding area. Thus the perception of color depends on an integrated visual activity.

This experiment demonstrates the phenomenon of 'simultaneous contrast.' The other principal type of visual contrast is the 'successive' one. It can be seen in the after-image, of which there are two types, the positive and the negative. Positive after-images can be produced when the eyes are in a state of rest, as, for example, after awakening in a dark room. If we look intently against a bright spot between the curtains for a moment, and then shut our eyes we shall find a bright after-image of the spot floating within the dark *eigengrau* visual field. This fades away rather quickly.

Negative after-images are more common, since they belong to normal viewing conditions. If we look intently for a moment at one particular colored area, which is then turned off, like a traffic light, for example, we may observe a negative after-image of the area in a complementary color.

Goethe studied these effects in detail. In the sentences quoted below he reviews the then developing receptor theory, which was later to become dominant within the science of vision, and which was indeed to play a great part in advancing our knowledge of the anatomy and the physiology of the eye. According to this theory, a receptor should be imagined as a tiny physico-physiological device which transforms patterns of electromagnetic energy into corresponding patterns of nerve activity. A crucial property of the receptor is its 'sensitivity,' which, according to this theory, plays more or less the same part as the sensitivity of a photographic film. The dark after-image of a bright spot might be explained as a result of a relative decrease in sensitivity of the corresponding part of the retina, the decrease being caused by the previous exposure to a relatively high light intensity. Goethe refers to this explanation and adds the following:

This method of explanation would appear sufficient for the point at issue, but in the light of later observations it is necessary to explain the phenomenon as originating from higher sources. The waking eye reveals itself as a living organism. This is shown by the fact that it categorically demands a change in its state of being. The eye is not capable of remaining—cannot bear to remain—for even a moment in a state which is determined by the object looked at. It is forced into a kind of opposition by setting the extreme up against the extreme, the intermediate against the intermediate; it unites opposites in a common bond. And both in succession and in congruity of time and place, it strives for completion[8].

Probably Goethe had no idea of the future success of the receptor theories. So much the more interesting to see him take a decisive stand on the receptor's function. The higher sources referred to must be the ultimate aim of the visual process, which is to construct a 'picture' of the world. This picture does not arise through a one-way process, as with photography, but through the interaction of an inner and an outer dimension. The visual picture must be described in dynamic terms. *The outer world is being projected into the organism, and the inner world is projected into the environment. Both processes are taking place through the action of the living eye.*

Two fundamental doctrines have been accepted in visual science during the last hundred years, and there have been certain problems of communication between them. There is a physicalistically directed thinking which considers the visual picture as a pointwise representation of the environment, where the process of representation actually consists of a series of physical actions. In this view one particular color represents a mixture of light rays. A closer examination of the laws of color mixture in connection with the spacing out of colors on a color circle reveals that color vision in this sense may operate by means of three primary colors or color systems. A more psychologically directed thinking looks for the lawful structure of the visual picture as such. A visual picture is built up from elements of its own, colors, for instance, with appropriate visual properties. Take, for instance, the so-called 'color circle,' which can be visualized in different shapes. The fact that hues can be systematically distributed along a closed curve cannot be derived from physical principles. The color circle appears as a result of the properties of colors themselves. For instance, a yellow color can without loss of chromaticity change into a red one, but not into a blue one.

These two opposing views within the science of vision have been connected with the names of Young and Helmholtz, on one side, and with the name of Hering, on the other. And there has been a tendency to overlook and reinterpret opposing views. The controversy has been much concerned with the function of the color receptors in the retina. Now it is an interesting fact that Thomas Young's original three-receptor hypothesis was not founded on anatomical and physiological findings, but partly on a reconsideration of the seen spectrum itself. Young was impressed by the dominance of three hues in the seen spectrum, namely, red, green and blue-violet, whereas Newton had counted seven full members of the spectral family. From his own observation Young concluded that the eye may operate by means of three types of color-sensitive nerves, responsible each of them for the aforesaid color sensations, and producing the rest of the colors through the sum of their interactions. This historical incident can teach us how difficult it is to draw a sharp line of division between the scientific attitudes mentioned above. The incident also points to the distinction between one particular, experimental or natural representation of color, like the spectrum, or the rainbow, and the ideal representation of hues in a color circle. In fact, Newton considered the spectrum as something like the color circle itself, and this misunderstanding has invaded many textbooks in physics till this day. Probably Goethe was the first who paid attention to the striking difference between the color structure of a spectrum and that of an ideal color circle. A suitable representation of the color circle making use of the primary colors in additive and subtractive color mixtures is the following:

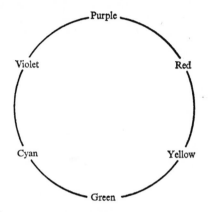

In the spectrum, purple (magenta) and the charmoisin-red transitions do not appear at all, and the yellow and cyan hues lack the full color quality of their red, green and violet neighbors. This striking property of the spectrum, noticed by Thomas Young, was mentioned by John Herschel, too, in his great treatise on light. But since then very few people have observed the spectrum with open eyes, and almost every colored reproduction of the spectrum in textbooks nowadays shows a brilliant yellow, which makes the picture beautiful, but not true.

Although Goethe's name is mentioned more often by the workers within the Hering school, and although Hering himself was much inspired by Goethe's work, it would not be justified to consider Goethe a real forerunner of Hering. Except for the few sentences already quoted, Goethe was not explicit on the question of the receptor functions. The basic features of the three-receptor theory were still little known at that time. Goethe's knowledge about the laws of color mixture were insufficient compared with the present state. It would therefore be tempting to project Goethe's ideas on this point.

Our first presupposition would be that the task of the eye is to produce a pictorial representation of the external environment, i.e. a picture where each object has been assigned the color that belongs to it under the momentary conditions. The eye might reasonably be expected to accomplish its aims if it were equipped with a set of reference bodies, the characteristic colors of which were suitably spaced out along the color circle. The eye then has to put these objects up against the outer objects for comparison. If this is so, then, what would be the least number of bodies necessary for the function of the eye? For reasons touched upon already, it follows that three bodies of reference might be enough. Between any two points on a closed curve, one can travel along two routes that can be distinguished by means of a third point of reference. And this hypothesis has been confirmed by the anatomical-physiological findings, if we regard the receptors themselves as such objects of reference. Our hypothesis coincides with that of Thomas Young, although it was arrived at from another point of view. In Young's hypothesis the receptors are the origin of color sensation, whereas we prefer to consider them as organs of correlation between the colored object and the living individual.

However, some basic function is still lacking in our model of the eye. We have been discussing color as part of the object, but we have not met with the seen color as such. The property of being seen links color primarily with the life of the individual: color appears as a purely visual entity. And again we may ask how many primarily different seen hues the eye must know about in order to be able to distinguish the hues along a color circle. A rather simple argument will convince us that three colors of reference will not do. Say that we recognize a certain hue F_1. As a member of the color circle, F_1 must be primarily distinguishable from a certain hue F_2 and a certain hue F_3, one on each side of F_1. But F_1 must also be primarily distinguishable from a certain hue F_4 somewhere on the opposite side of the circle. In other words, a seen color circle must contain at least four primarily distinguishable hues. And in fact these four unique hues of the color circle have been known since Leonardo as yellow, red, blue and green. Other members of the color family appear as transitions, like orange, for instance, which is a yellow-red color.

So much about the systematics of seen colors. Finally, we have to connect the phenomenon of the seen color with the act of vision. The seen color has an active quality that is rarely mentioned because it is hard to express in scientific terms, but everybody who is a little trained with colors knows for himself that color has an active, living quality. The thought of losing contact with seen color is felt by the individual as something like the thought of losing one's arm. What concerned Goethe in his treatment of physiological colors was the idea that color itself is the proper organ for the act of vision. The act of vision really is an act of visualizing, i.e. making visible something that is already there. A picture is not a creation by the individual, nor by the object. For centuries science has looked for the origin of pictures in the brain and in the object, but in vain. The rise of a picture can only be explained by something that is itself already there, in this case the color. But, as we have shown already, to connect the visualizing process with the external object is brain work. The brain, considered here as the integrated sum of receptor functions, operates as an organ of correlation between the visualizing individual and its environment. But the visualizing process itself cannot be reduced to brain functions; it can only be brought back to the existence of colors. This, I think, would be the standpoint of Goethe as compared with some recent ideas in the science of vision.

In his experiments with physiological colors, Goethe was much concerned with the phenomenon of color contrast because it seemed to him to reveal the active quality of seen color. He explained the picture process as the state of equilibrium between opposing colors. This rough idea of his has to be taken in an intuitive sense. It should be remembered that the phenomenon of color contrast is an intriguing and unsolved problem in modern visual science. Honor is due to Goethe for having recognized the phenomenon from a higher point of view.

There exist some interesting experiments which show the intermingling of those two structural principles of the visual cosmos, the trichromatic and the opposing-color one. Among these should be mentioned Dr. Land's 'two color projection' experiments.[9] They have met with a certain scepticism, since they seem not to have shown anything 'new.' Dr. Land's effects *can* be considered as 'simultaneous contrast.' Still the important question that arises from Dr. Land's experiments is this: How do we explain 'simultaneous contrast' from a higher point of view?

Dr. Land's experiments are not easy to perform. They

demand proper photographic work. But the following experiment, which is easily set up, will show the intermingling of the two above-mentioned structural principles. With three slide projectors, we produce on a screen three partly overlapping dark areas within a bright field. If we now place before the projectors color filters, such as green, red and violet, we can observe on the screen a demonstration of the principle of trichromacy according to the formula:

$$Green + Red = Yellow$$
$$Red + Violet = Purple$$
$$Violet + Green = Cyan$$

But if we remove, say, the red filter, we observe to our surprise that the presumed grey area still appears reddish. The effect can be increased by using, instead of violet and green, a yellow-green and a blue-green filter. Now the respective light fields appear in blue and yellow, whereas the presumed grey area appears with a magenta color that seems at least as saturated as the two others.

Thus we have been able to produce a total color gamut by mixing appropriately two chromatic hues + white light, although the former experiment, which uses a set of three chromatic 'primaries,' is the only experiment mentioned by all the common textbooks. If we compare our impressions of those two-color mixture experiments, we are struck by the fact that the color gamut produced by the 'two-color mixture' appears much better balanced from an aesthetic point of view, while the three-color gamut does not satisfy our feeling of color harmony. This very striking fact has a natural explanation. When only two primaries are offered in a suitable mixture, the eye is more free to operate according to its own intrinsic laws, while in the three primary mixture, conditions for color appearance are relatively fixed. This is the reason why we preferred to perform the color mixtures by overlapping *dark fields with a light surround*, although it has long been common practice to operate with light areas on a dark background. Of course, the two-color effect will appear upon a dark background also, but not with the same brilliancy. To the best of my knowledge, the difference between these mixture experiments has not previously been mentioned or seriously discussed.

The dark field procedure may help explain a rather obscure Goethean sentence, namely, that 'color is something shadowy.' Color appears when there is the proper equilibrium between the eye and the field. Then, as we have seen, the darker shades of the visual picture can be considered as the result of the activity of the eye itself. Therefore also, within those dark shades of the visual area, color may appear with its greatest brilliancy.

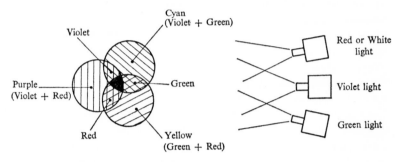

White surround (Violet + Red + Green)

The physical colors. The 'physiological' colors take the first place as objects of study because of their fluctuating character, which calls our attention to color as something emerging. Nevertheless, the 'physical' colors have a central position in the progress of Goethe's investigation. Whereas the physiological chapter has been universally acclaimed as the work of a pioneer by all kinds of opinions, the physical chapter has aroused much controversy. The ideas of the latter chapter were rejected as un-scientific by Helmholtz; they were violently attacked with experimental arguments by Ostwald; and they were held questionable in philosophical and experimental arguments by Schopenhauer, who nevertheless regarded himself as a true defender and pupil of Goethe. Modern and positive commentators, such as Heisenberg, Weizsäcker, Heitler and others, admit their uncertainty about Goethe's ideas on 'physical colors.'

The reason for this is that Goethe deals with light as something undivided and invisible that reveals itself in an *Urphänomen*, i.e. a complete set of singular phenomena, while in the Newtonian view white light is considered a mixture of colored lights (or in other words: color-impressions producing lights). And yet if one discusses Goethe's *Farbenlehre* with painters,

63

one does not meet with the same terminological difficulties. In the painter's view, too, light is invisible and undivided. What concerns the painter is the activity of light, light as the dynamic structure of a picture. Goethe *speaks* about light in much the same way as painters do, but he *deals* with it experimentally as physicists do. Does he then deal systematically with nature as a picture? Yes, he does. His experiments are methodical attempts concerning the appearance of nature as a picture. In other words, they are questions about the laws of coloring. Both Goethe and Newton are occupied by the appearance of the 'spectrum,' but from different points of view. Newton, the atomist, develops the idea that the spectrum consists of separate kinds of light rays. Goethe, the morphologist, recognizes a picture, and he looks for its inner dynamics.

Of course, we must admit that we are now using the word 'picture' in a different sense from that which has become customary in physics since Kepler. The Keplerian 'picture' is a purely geometrical proposition, which states that when two or more rays of light from the same point having passed through some optical device meet again in another point, this second point is a 'picture' of the first. Kepler's definition runs in terms of geometry, Goethe's idea in terms of light and darkness. It may be objected that the Keplerian conception has the advantage of being defined in exact and objective terms, while Goethe never tells us explicitly what a picture really *is*. He only points to natural categories of pictures. He acts like a painter, who does not occupy himself with discussing what a picture *is*, as his aim is to *create* pictures. Goethe's aim is to *see* the picture. The picture is the higher phenomenon within the phenomenon. Goethe's *Farbenlehre* obviously takes up an old tradition in natural philosophy, which is to exhibit nature as the great Book of Pictures. About the time of the *Farbenlehre* there appeared two great works that can be compared with it in so far as they, too, were motivated by this joyful contemplation of natural things—namely, the treatises on light and color by Joseph Priestley and by John Herschel, the one a renowned chemist, the other a world-famous astronomer. Of these two, Herschel especially is influenced by the Galilean idea that 'the great Book of Nature is written in mathematical letters.' Herschel therefore concludes with a brilliant piece of mathematical optics. Goethe's idea is that the picture is a product of man's relations with nature, and is the highest expression of nature.

We cannot attempt to reproduce here the richness of the Goethean treatment. We can only indicate certain general features of Goethe's exposition of 'physical' color phenomena. 'Physical' colors occur when light shines upon, into, or through certain more or less translucent materials which are themselves colorless. These phenomena are subdivided into several classes. The colors of the atmosphere are 'physical,' as are those of the rainbow, of Newton's spectrum, of the gleaming diamond, of oily films on street puddles, of the surface of hardened iron, etc. In the physicist's language the physical colors are those of scattering, refraction, interference, diffraction, polarization, etc.

In the *Urphänomen* of this class, color occurs as a transition between light and dark, either as lightened darkness or as darkened lightness. In either case color can be considered something like a 'half-shadow.'

The *Urphänomen* of color appearance is clearly demonstrated in the atmosphere. On a bright, cloudless day, the polarity of light and darkness produces the duality of the whitish-yellow disc of the sun and the blue sky. The yellow thus appears as the color which comes closest to light itself; it appears as 'darkened light,' whereas the blue of the sky appears as 'lightened darkness,' namely the darkness of the universe. This fundamental polarity gives rise to a multitude of atmospheric color situations, as the conditions of the atmosphere and the position of the sun change. As the sun sinks toward the horizon, a wider layer of atmosphere fills the space between us and the sun. Accordingly the light of the sun appears more and more darkened. The sun takes on a reddish hue, which under certain conditions can turn into a dully glowing ruby red. Thus we observe one primary color transformation that is brought about by the increasing darkening of the light. But simultaneously, as the blue sky darkens, it deepens to darkish blue and even violet before it takes on the black color of the night. We are obviously witnessing a duality of darkening processes, on the one hand, a set of colors that are related to the light of the sun, on the other, those colors related to the darkness of the universe.

But what about the rest of the color continuum, symbolized

65

in the color circle? The green and purple? In fact, there are no processes in nature that produce green and purple through a further darkening of the light or a lightening of the darkness. Nevertheless, green and purple can be observed in the atmosphere, namely as *colors of culmination.* During sunset we frequently observe a yellow-green region of transition between the yellowish regions around the sun and the bluish regions of the evening sky. We can also see rich pinks, and even purple appears as a transitional color on the evening sky, as, for instance, when clouds seen against the dark sky receive the red evening light from the setting sun. The appearance of green as a color of

Newton's principal arrangement.

culmination is confirmed by another brilliant phenomenon: the rainbow, where the green band divides the red/yellow from the violet/blue ones.

The prismatic colors. These phenomena were studied in great detail by Goethe because they threw light upon the dynamic structure of the color circle. They also supplied his principal experimental arguments against Newton.

The difference between the Goethean and the Newtonian approach is obvious. They each started with what they considered to be the simplest conditions for their experiments. For Newton this was the 'ray of light,' which he produced by letting the light from the sun enter his room through a narrow hole in the window shutter. The 'ray of light' was then passed through a prism, where it spread out and produced a 'spectrum' on the opposite wall.

Goethe, on his side, looked through the prism at the simplest picture he knew. Next to the neutral, boundariless surface, the simplest of all pictures may be said to be that which consists of two shades only, for instance, a black and a white area separated by a straight line. Seen through the angular prism, the picture will appear displaced according to a rather simple law, and the border-line between the white and the black areas will appear as a colored transition. Everyone doing this experiment for himself will soon recognize that, in order to produce the optimal color effect, he must arrange conditions such that the straight line is displaced along a direction vertical to the line itself. Then the border-line will either be displaced toward the dark part of the picture, or it will be displaced toward the light part. These two opposite conditions lead to a duality of color phenomena. If the displacement occurs toward the light area, the transition consists of yellow and red hues, according to the following pattern:

Black–Red–Orange–Yellow–White.

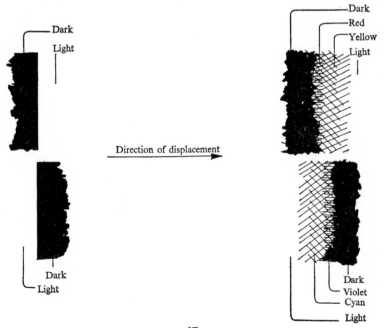

If the displacement occurs toward the dark area, the blue and violet hues arise:

White–Cyan–Blue–Violet–Black.

In both cases the saturation of the colors grows with increasing darkening.

These effects, as well as the following ones, can be demonstrated objectively with the help of a slide-projector passing corresponding slides through a glass prism. The boundary colors will appear in great brilliance on the screen.

Our first observation was similar to that with the atmospheric colors. A certain development of color took place between light and darkness. So we have to ask ourselves again: How must we alter the experimental conditions in order to produce those parts of the total color continuum which have not yet appeared?

We aim at an interaction between the complementary phenomena that we have already produced. Let us therefore look through the prism at the simple pictures consisting of *two* lines *enclosing* respectively the darker and lighter shades. As these two lines approach one another, the corresponding colored fringes that are produced by the prism will partly overlap. In order to have a gradually increasing amount of overlapping, we tilt the lines so that we are looking at wedge-shaped figures. In the middle of the *light* wedge, where the yellow and blue bands must overlap, green occurs, yellow and blue *darkening* each other. The top of the wedge now exhibits the ordinary spectrum, with its three dominant bands: violet, green, and red.

In the middle of the *dark* wedge, however, where the red and the violet bands must overlap, a brilliant purple occurs, as the result of a *lightening* process. The red band lightens the violet. The final result of the color interaction along the dark wedge is thus a second spectrum where the dominant color bands are yellow, purple and cyan. The latter spectrum is the dual counterpart of the Newtonian spectrum which was formed on the top of the light wedge.

Of course these experiments may be continued. We may change the conditions, using colored lights, and more complicated pictures, etc. But we shall not go into detail here, since the fundamental fact has been demonstrated, namely, that it is possible to derive the colored appearance of a spectrum from general relations between colors.

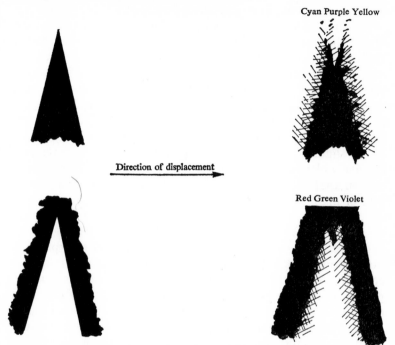

Cyan Purple Yellow

Direction of displacement

Red Green Violet

The chemical colors. Goethe's treatment of the 'chemical' colors, that is, those colors which belong primarily to the object itself in virtue of its exterior appearance, is rather a rough draft. The reason for this may be found in the circumstance that here the richness of the phenomena becomes overwhelming. The physiological and physical phenomena could be derived from simple and general ideas; but in the chemical category apparently everything determines its own color-cycle. Thus the nature of the *Urphänomen* becomes obscure.

Goethe obviously recognizes in the changing of colors a reflection of the life of the thing itself. He is not primarily concerned with the momentary color of the thing. What he asks is this: How does this particular color appear and disappear? Referring to the totality of the color circle, he aims at a general description of the dynamic elements of these changes in the colored substances. The color movements are classified according to principles of transformation along the color circle, as when he speaks about color 'appearance,' 'development,' 'culmination,' 'fixation' and 'disappearance.'

A typical Goethean view is revealed in an interesting sentence about the nature of color in relation to the natural hierarchy. Goethe, the morphologist, had the idea that the higher the ranking of the organism, the higher is the degree of its freedom in relation to environmental influences, and that this freedom is made possible by the organism's mastering a wide range of functions. Thus the color of the surfaces of the higher organisms is more complex, just as with the human skin, which is sensitive to the slightest stirrings of the inner life. Goethe also refers to the lively elementary colors which adorn parts of the bodies of certain apes as proof of their primitive nature as compared with that of man.

CONCLUDING REMARKS

We have merely touched on a vast subject that is surprisingly little known today. Several questions that arise as a matter of course have had to be by-passed. One of them concerns Goethe's position within the history of natural philosophy. There is an obvious relation between the color theory of Goethe and that of Plato. This is so much the more interesting as Goethe's treatment is strictly empirical and experimental, while Plato's forms part of a philosophical cosmology.

Neither did we engage in a discussion of questionable statements which are naturally to be found in an ambitious work of this kind. The primary outcome of wandering in Goethe's territory is educational. The occupation with color appearance from the higher point of view may contribute to the maturation of the human personality. Natural phenomena exhibit themselves with a new richness whenever the individual is confronted with the realm of pure experience. He recognizes the limited scope of every formula. But in the light of pure experience the formulae become more meaningful. Words like 'light,' 'darkness' and 'color' become deeper and richer words.

Goethe's color theory should be seriously reconsidered if only for its pedagogical value, as it is in fact the only modern approach to the realm of visual experience from a unifying point of view.

REFERENCES

[1] There are several editions of Goethe's color theory. Among these might be mentioned: *Werke*, Hamburger Ausgabe, *13* (1960), which contains a large bibliography; dtv-Ausgabe, *40–42*, pocket edition. Of Goethe's general scientific writings, the following essays might be referred to as relevant to our statements: 'Der Versuch als Vermittler von Objekt und Subjekt,' 'Inwiefern die Idee: Schönheit sei Vollkommenheit mit Freiheit, auf organische Naturen angewendet werden könnte,' 'Erfahrung und Wissenschaft.'

[2] See, for instance, W. Heisenberg, *Wandlungen in den Grundlagen der Naturwissenschaften*, Leipzig: S. Hirgel, 1935; W. Heitler, *Der Mensch und die Naturwissenschaftliche Erkenntnis*, Braunschweig: F. Vieweg, 1961; trl. *Man and Science*, New York: Basic Books, 1963, Chap. 2.

[3] H. Helmholtz, *Physiologische Optik*, Leipzig, 1867, p. 268.

[4] Goethe's correspondence with Eckermann; December 21, 1831.

[5] Goethe to Schiller; November 15, 1796.

[6] From *Sprüche in Prosa*.

[7] From the essay, 'Erfahrung und Wissenschaft.'

[8] *Farbenlehre*, sec. 32–35.

[9] E. H. Land, 'Color Vision and the Natural Image,' *Proceedings of the National Academy of Sciences*, *45* (1959), Pt. 1, pp. 115–29; Pt. 2, pp. 636–44.

3

IS BIOLOGY A MOLECULAR SCIENCE

Barry Commoner

Biology is concerned with the elucidation of the remarkable properties which distinguish living things from all other forms of matter. Since the living cell is composed of substances which are fundamentally similar to those not associated with life, this problem takes a specific form: Are the cell's unique properties, at least in principle, predictable from the observed properties of its isolated molecular components, or are they due to some property which is not discernible in these separate constituents?

That such a question is burdened with profound theoretical and practical consequences is readily apparent from an analogous instance—the atom—in which the answer is already clear. The unique properties of an atom—such as its stability and its capability for combination with other atoms—are predictable from the properties of certain sub-atomic constituents, which are readily observed in isolation: the electron, the proton, and the neutron. Detailed knowledge of the separate properties of these atomic components, and of their various modes of interaction, have yielded penetrating insights into the properties of the whole atom. Can we anticipate corresponding results from an analysis of the relationship between the properties of the separate cellular components, and the unique attributes of the whole cell?

It should be stated at the outset that if scientific questions were subject to the will of the majority, then the answer to this question has already been given. Most practitioners of biology and its attendant sciences seem to be convinced at this time that all of the unique attributes observed in a living cell are the discernible consequences of known, or at least knowable, properties of the cell's molecular constituents. In particular, there is a widely held view that those special properties which

73

are most fundamental to life—growth, replication, and inheritance—arise from the distinctive properties of a particular constituent universally present in living things, deoxyribonucleic acid, or DNA. This conviction, added to earlier evidence that other properties of living things, for example, metabolism, are determined by the chemical properties of specific cellular constituents, such as enzymes, leads to the more general conclusion that biology is, fundamentally, a molecular science.

But momentary popularity is not always the best indicator of the validity of a scientific theory (witness the phlogiston theory), and it is useful to examine the now widely accepted conclusion that biology is a molecular science against specific, objective criteria. The best test of a concept is, of course, its inherent necessity from the relevant scientific data. If it can be shown, for example, that the inherent properties of certain cellular constituents are *necessary and sufficient* to explain a unique cellular attribute, then biology is indeed, in this area at least, a molecular science. Because of the central importance of inheritance and self-duplication in the unique attributes of life, it is especially useful to determine whether the known properties of DNA are necessary and sufficient to explain inheritance, as exhibited by living organisms.

Inheritance involves the reappearance in the offspring of certain specific transmissible features of the parent. In the simplest case, a single cell on division gives rise to two daughter cells, both of which exhibit the inheritable specificity inherent in the original parent cell. This is the process of *self-duplication*. Although self-duplication of the cell can occur only in a suitable system (for example, one containing necessary nutrients), the inheritable characteristics of the progeny are generally *not* influenced by the environment, and the specificity of the new cell is derived only from the specificity of the parent cell. In this system (i.e., the cell in a suitable nutrient environment), we may refer to the original cell as a *regulatory* factor (in that it controls the specificity of the progeny), its participation in the process being indicated by the symbol ($\Longrightarrow$). The participation of any other component of the system, such as a nutrient, which is essential for the process but does *not* regulate the specificity of the product, may be symbolized by ($\longrightarrow$). Hence, in the case of a self-duplicating cell, we have the relationship:

original cell $\Longrightarrow$ new cell

nutrients, etc.

Where only *one* of the necessary components in a system is regulatory, we may refer to it as a *germinal* component, indicated by the symbol ($\Longrightarrow$). Hence, in the foregoing case:

original cell $\Longrightarrow$ new cell.

This then is the definition of *self*-duplication: *The specificity of an organism or a cell is determined only by its own specificity; the organism, or the cell, is a germinal component.*

Having at hand this means of generalizing the relationships involved in self-duplication (see Fig. 1), we can apply the foregoing concepts to the molecular events which, according to the DNA theory, are believed to be the source of the cell's unique capability for self-duplication.

To begin with, we need to translate the meaning of biological specificity to molecular, or biochemical, terms. There is at this time a considerable body of evidence which shows that at least some specific inheritable properties of living organisms are due to specific attributes of a particular molecule. Thus, the inherited color of a flower is due to the presence of particular molecular pigment in the cells of the flower, and this is, in turn, a consequence of the capability of the cell's chemical systems to synthesize this specific pigment. Hence, in at least certain instances, a specific feature of inheritable biological specificity can be reduced to the specificity of the cell's system of biochemistry, that is, to *biochemical* specificity. Specificity may be regarded as roughly equivalent to 'information content,' although the relationship of the latter term, as applied to biochemical and biological systems, to classical information theory is by no means clear.

Biochemical specificity falls into two general classes:

(*a*) *Static* biochemical specificity is represented by the particular order of the residues which comprise linear biological polymers. In such molecules, a very large number of primary structures (i.e., sequences of residues) have equal thermodynamic probability, and the static specificity of a polymer represents the specification of a particular sequence among the many possible ones. Thus, the biochemical specificity of a

DETERMINATION OF SPECIFICITY

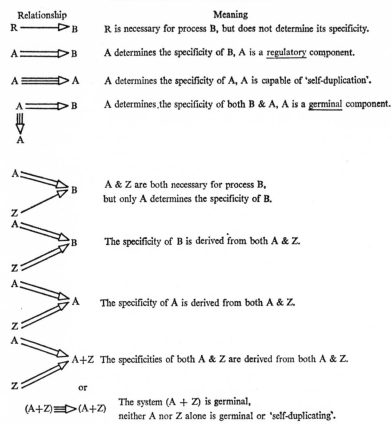

Relationship	Meaning
R ⟶▷ B	R is necessary for process B, but does not determine its specificity.
A ⟹▷ B	A determines the specificity of B, A is a regulatory component.
A ⟹▷ A	A determines the specificity of A, A is capable of 'self-duplication'.
A ⟹▷ B	A determines the specificity of both B & A, A is a germinal component.

A & Z are both necessary for process B, but only A determines the specificity of B.

The specificity of B is derived from both A & Z.

The specificity of A is derived from both A & Z.

A+Z The specificities of both A & Z are derived from both A & Z.

or

$(A+Z) \Rightarrow (A+Z)$

The system (A + Z) is germinal, neither A nor Z alone is germinal or 'self-duplicating'.

FIG. 1. Diagram to summarize definitions of symbols for non-regulatory participants, regulatory and germinal components of systems involving transfer of biochemical specificity.

protein is represented by the precise sequence of the twenty-odd different amino acids that can occur in this linear polymer. Similarly, the biochemical specificity of a nucleic acid is represented by the sequence in which its four or more constituent nucleotides form the linear array of this polymer.

(b) *Kinetic* biochemical specificity is represented by a particular path of chemical reactions in the cell (for example, the sequence of reactions that leads to the synthesis of a particular pigment). The numerous molecular components of the cell are thermodynamically capable of a very large number of reactions, and the specification of a particular metabolic pathway represents the activation of relatively few reactions among the numerous possible ones.

The establishment of either static or kinetic biochemical specificity requires the regulation of the rates of the relevant intracellular reactions. Such regulation will determine which of the alternative reactions actually occur at appreciable rates in the cell, and thereby determines the outcome of cellular chemical events.

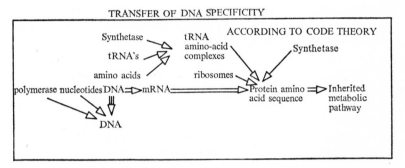

FIG. 2A. This diagram illustrates the transfer of biochemical specificity in the system proposed by the theory of the DNA code, which holds that the biochemical specificity of DNA (its nucleotide sequence) is the sole source of specificity in the replication of DNA and in the synthesis of protein, the specificity of the latter determining a particular inherited metabolic pathway. In this scheme, DNA is itself a germinal self-duplicating component and the sole source of the cell's inherited biochemical pattern.

The well-known theory that DNA governs biological inheritance holds that the static specificity of its molecular structure—the sequence of nucleotides in a particular segment of DNA—is the sole source of both the specificity of further DNA synthesis, and of the kinetic specificity of a particular metabolic reaction in the cell. The theory proposes the stepwise transfer of the static specificity embodied in the nucleotide sequence of the DNA to the amino-acid sequence of a protein of a particular enzyme by a series of molecular processes which are summarized in Fig. 2A, using the symbols defined earlier. The scheme

77

leads to the notion that DNA is a 'self-duplicating molecule,' because it holds that DNA is the only source of the specificity of newly synthesized DNA. Moreover, according to this scheme, the biochemical specificity of a given protein, which is due to its amino-acid sequence, arises ultimately from a single source— the nucleotide sequence of a particular DNA segment. Since all enzymes are composed, at least in part, of proteins and regulate the kinetic specificity of the cell, this system offers an explanation of *biological* self-duplication and inheritance.

According to this scheme, a specific nucleotide sequence in a DNA molecule has the attributes of a gene, in that it regulates its own replication (being thereby transmissible to the progeny) and also regulates the biochemical specificity which is the source of the organism's specific biological characteristics. Although, as will be shown below, the gene does not exhaustively account for the origin and transmission of all of the inherited features of an organism, it certainly does account for those features described by Mendelian genetics and therefore merits an effort to explain its molecular origin.

The mechanism which is supposed to account for the capability of DNA for self-duplication was first proposed by Watson and Crick. According to them a single strand of DNA attracts to each of its constituent nucleotides (which we may abbreviate as A, T, G, and C) a specific complementary nucleotide, the association being mediated by weak hydrogen bonds. Thus A attracts T, T attracts A, C attracts G, and G attracts C. By this means, the original DNA fiber causes a parallel alignment of a specific sequence of nucleotides. It is proposed that chemical bonds are then formed between successive nucleotides by DNA polymerase enzyme, thus forming a new DNA fiber on the 'template' provided by the original DNA fiber. According to this scheme, then, the regulatory relationships are:

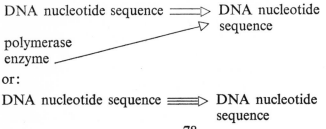

or:

DNA nucleotide sequence $\Longrightarrow$ DNA nucleotide
                                            sequence

It is proposed further that the DNA nucleotide sequence is the sole regulator of the nucleotide sequence of a 'messenger' ribonucleic acid (or mRNA) and that the latter is, in turn, the sole regulator of the amino-acid sequence of enzymatic protein, thus determining the kinetic specificity of the cell. Hence: DNA nucleotide sequence⟹mRNA nucleotide sequence⟹ protein amino-acid sequence⟹enzyme specificity⟹ kinetic specificity of the cell.

The foregoing conclusions are valid only if it can be shown that no other *regulatory* components participate in DNA synthesis or in the processes which intervene between the DNA 'template' and the synthesis of a particular protein. It becomes important, therefore, that the experimental evidence be scrutinized with these questions in mind.

Three classes of components are operative in DNA synthesis *in vitro*: the DNA primer or 'template,' the enzyme which catalyzes the synthesis of DNA (DNA polymerase), and the deoxyribonucleotides which are necessary to form the DNA polymer. That the biochemical specificity of the newly formed DNA (i.e., its nucleotide sequence) is under the *partial* control of the primer is evident from a number of Kornberg's observations of similarities in the composition of primer and product DNA.[1] However, the correspondence between the nucleotide composition of primer and product is not always complete,[2] and in any case no present method yields more than very scattered data regarding nucleotide sequence. Moreover, it has not yet been possible to demonstrate the actual replication of a biologically active DNA.[3] Thus, evidence from synthesis of DNA *in vitro* shows that the primer DNA is a *regulatory* component and influences the biochemical specificity of the DNA product. But there is no direct evidence that the primer exerts *total* control over the specificity of the product, i.e., that it is a *germinal* component and therefore capable of 'self-duplication.'

On the other hand, certain *in vitro* experiments also show that the polymerase enzyme is itself capable of regulating the specificity of the DNA product. When the polymerase is added to a mixture of deoxyadenylic and deoxythymidilic acids, *in the absence of a DNA primer*, DNA synthesis is observed after a lag period.[4] The product is an A–T copolymer with a regular

79

alternating sequence of nucleotides: ... ATATAT ... When a corresponding experiment is performed with a mixture of deoxyguanylic and deoxycytidilic acids, two polynucleotide products are formed: ... GGGG ... and ... CCCC ... In these cases each of the DNA products has a regular and quite specific, albeit simple, nucleotide sequence and, therefore, possesses biochemical specificity. Since primer DNA is absent, this specificity must originate in the enzyme, which, in the absence of a primer, appears to catalyze specifically certain internucleotide bonds.

Recent work from Kornberg's laboratory provides us with a direct test of the template hypothesis of DNA synthesis, on which the notion of DNA 'self-duplication' rests.[5] In these experiments a highly purified sample, consisting of a nearly homogeneous preparation of DNA fibers about 11 microns long, was used as a primer in the DNA polymerase reaction. The product was characterized by electron microscopy. It was a fiber much longer than the primer and highly branched; while the original fiber had only two ends, the product had an average of twelve ends. Clearly this result is not indicative of a process of self-duplication. (See note c.)

More recent experiments by Khorana *et al.* cast further doubt on the self-sufficiency of the DNA template as the sole source of DNA specificity. In these experiments short lengths of artificially synthesized DNA (containing, for example, 7–11 nucleotides forming an alternating A–T sequence) are employed as primers in the conventional DNA polymerase system.[6] The product also exhibits the alternating A–T sequence of the primer, but is very much longer; it has a molecular weight ranging into the millions. Such a result cannot be achieved by a simple side-to-side template process. At best it requires that the long-chain product be determined by a series of laterally displaced short-chain templates; but in this case a new mechanism is required to regulate the proper relationship between the positions of successive templates. At worst (for the template theory) this result means that replication does not involve a template at all, the new polynucleotide being added terminally to the primer; in this case, the formation of the hydrogen-bonded complementary strands which are observed in the product may be due to a secondary crystallization not at all

80

related to the replication process itself. Significantly, Khorana also finds that the synthesis of a long-chain DNA is greatly enhanced when a short chain of identical nucleotide sequence is present in addition to the template, the new chain apparently adding terminally to the short one.

A recent *in vivo* experiment leads to the same conclusion.[7] Here it was observed that mutations of the bacterium *E. coli* which alter its DNA polymerase enzyme cause a high frequency of specific mutations in other genes. If, as widely believed, the gene is represented by a segment of DNA with a specific nucleotide sequence, this result means that the enzyme which synthesizes DNA can determine, in part, the nucleotide sequence of the DNA which it produces. In the words of the authors: 'this indicates that the polymerase is involved in base selection during DNA replication.'

Hence, according to the present data, precise replication of DNA nucleotide sequence requires the ordered interaction among at least two molecular processes: (i) the influence of the nucleotide sequence of a pre-existing DNA fiber on the nucleotide sequence of a new DNA fiber synthesized in the presence of the original one, (ii) the specificity of the polymerase for the formation of particular internucleotide bonds during the synthesis of DNA catalyzed by this enzyme. The simple template-copying originally proposed in the Watson-Crick scheme is clearly inadequate to explain these relationships.

There is no known mechanism, apart from the unknown one which exists in the intact cell itself, which provides a specific coordination of these elements sufficient to ensure precise replication of a complex DNA fiber, nor is there any good evidence that this has yet been achieved *in vitro*. (See note c.) It cannot be said, therefore, that precise replication of DNA, which is the hallmark of biological reproduction, is due solely to the inherent chemical capabilities of the DNA molecule.

Another essential element of the view that reproduction is accountable for by the observed properties of isolated molecular constituents is that the biochemical specificity of a protein (i.e., its catalytic specificity as an enzyme) can be traced solely to the biochemical specificity (i.e., nucleotide sequence) of a messenger RNA, which is itself determined by the nucleotide sequence of a complementary DNA template. However, the available data

81

fail to support this conclusion. Thus it has been shown that the precision with which the messenger's specificity is translated into protein amino-acid sequence is strongly affected by the biological origin of an enzyme, RNA amino-acid synthetase, which participates in the process of protein synthesis. Several recent experiments also show that the 'code' which is supposed to convert messenger RNA nucleotide sequence to protein amino-acid sequence is not inherently precise in *in vitro* systems.[8] Marked changes in the influence of messenger RNA nucleotide sequence on the incorporation of specific amino acids into protein are observed when the pH, the Mg^{++} concentration, or the temperature of the system is altered.[9] These results mean that *in the cell* there must be some regulatory effects, other than that embodied in messenger RNA, to ensure the observed precision of *biological* protein synthesis in the face of environmental variations.

Some recent observations suggest, on their face, that the specificity of messenger RNA is indeed in full control of the specificity of protein synthesis. For example, a recent paper from Zinder's laboratory claims that the specific coat protein of a bacterial virus is synthesized *in vitro* when an algal protein-synthesizing system is influenced by virus RNA (serving as messenger).[10] What the *data* show is the following: Isotopic analysis of tryptic digests or protein synthesized in the heterologous system (algal enzymes and virus RNA) shows that several protein fragments that are characteristic of virus protein are identifiable. However, the same radioautographs also show that a number of *other* peptides characteristically produced in algal extracts are different from those produced in *E. coli* extracts. Hence, the data require us to conclude that when bacterial virus RNA is added to an algal system, the synthesis of new protein is not precisely determined by the specificity of the RNA; some protein fragments appear to be similar (though not necessarily identical to those of virus protein, for the internal amino-acid sequences of the peptides have apparently not been determined) to those of virus protein, but the new protein contains other amino-acid sequences as well and is therefore not identical with virus protein. Thus, although the messenger RNA clearly *participates* in the determination of protein specificity, its influence is by no means sufficient to account for

the remarkable precision of protein synstheis in the actual
biological process. We must conclude, therefore, that in the
intact cell some specific coordination of the influence of
messenger RNA, or RNA amino-acid synthetase, and of other
as yet unknown factors must account for the observed precision
of biological protein synthesis.

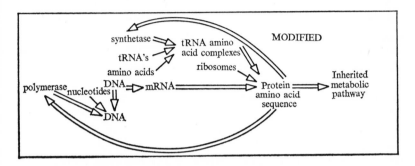

FIG. 2B. This diagram illustrates modification of the system
suggested by the evidence discussed in the text. This scheme
illustrates the consequences of evidence that part of the
biochemical specificity of DNA originates in the polymerase
enzyme which catalyzes DNA synthesis. Also shown is
the effect of contribution of specificity originating in
amino-acyl RNA synthetase, an enzyme essential to protein
synthesis. In this scheme DNA is no longer the sole source
of biochemical specificity for either DNA or protein synthesis; it is neither a germinal, self-duplicating component,
nor the exclusive determinant of the cell's biochemical specificity. Instead, these effects have a multi-molecular origin
and the entire system requires pre-existing specificity in both
DNA and at least two protein enzymes. Symbols according to
definitions given in Fig. 1.

Figure 2B shows how the additional sources of biochemical
specificity indicated by the evidence just brought forward
affect the conventional system. DNA specificity is no longer
wholly self-determined, and the specificity of a protein is determined not only by DNA nucleotide sequence (transmitted via
'messenger' RNA), but also on the species-specificity of the
synthetase.

Accordingly, the systems which govern both the specificity
of protein synthesis and of DNA synthesis are fundamentally

circular in nature. Although protein specificity is derived, in part, from DNA specificity, the latter is, in turn, partially dependent on the protein specificity of the DNA polymerase, which is itself dependent on the specificity of at least one other protein, the amino-acyl RNA synthetase. The system is operative as a self-sufficient source of specificity for DNA and for proteins only if DNA with a specific nucleotide sequence, and at least two proteins which possess species-specificity (polymerase and synthetase), are already present. This situation prevails in the intact cell. It can apparently be imitated with varying degrees of success in *in vitro* systems which derive all of the necessary components from the same biological source and thereby retain at least some of the complex sources of specificity which are present in the living cells of a given species. As indicated by a comparison of Figs. 2A and 2B, the net effect of the introduction of these additional sources of biochemical specificity is to negate the simple conclusion that DNA is by itself a germinal, self-duplicating component or that DNA is the sole source of the biochemical specificity, embodied in proteins, that regulates the specificity of cellular metabolic processes.

A number of experiments show that the introduction of DNA (e.g., bacterial transforming agents) into an appropriate host cell can induce in the latter the appearance of a new operational genetic agent.[11] This result does not conflict with the foregoing conclusion. A given DNA transforming agent is effective only when introduced into a living cell *of the appropriate type*. Hence, replication of the agent may depend on specificity contributed by the host cell.

Nevertheless, it remains true that, thus far, only DNA can serve as an agent of experimental genetic transformation. If, as suggested, proteins share with DNA an ability to determine inheritable biochemical specificity, should it not be possible to achieve genetic transformations with *proteins* rather than with DNA? Yet such effects have not been observed, and it is necessary to account for this fact.

The evidence of genetics shows that DNA agents which are the carriers of typical Mendelian effects are present in the cell in only one or two replicas; i.e., they are non-redundant. Hence, an invading DNA agent, such as transforming DNA, can readily compete with the DNA-borne genetic agents pre-existing in the host cell or, failing that, appear in an appropriate proportion of subsequent daughter cells unaccompanied by the indigenous agents. Thus the striking genetic effect of an invading DNA agent is due not only to its regulatory capability, but also to the low level of multiplicity of the cell's normal DNA-borne genetic agents with which it must interact.

However, in the case of a protein, for example, polymerase, matters are very different. A single cell probably contains hundreds or more separate molecules of a particular enzyme. For this reason, despite its postulated ability to contribute to the specificity of DNA synthesis, an invading polymerase molecule, with a specificity differing from that of the host polymerase, is not likely to have a noticeable effect on the biosynthetic activity of the invaded host cell and has little chance of becoming segregated out as the dominant form of the polymerase population of any daughter cells. Hence, no genetic effects are discernible. Nevertheless, such an observation does *not* prove that the polymerase is incapable of contributing to the specificity of DNA synthesis.

Thus DNA is uniquely capable of transmitting discernible genetic effects because of its unique lack of redundancy in the cell—and not because it is the only molecular constituent of the cell which governs biochemical specificity. Indeed, these considerations suggest that it might be possible to achieve genetic transformations with a *protein* agent—providing that some means could be found for artificially reducing the normally high level of redundancy of proteins in the cell.

The data of 'molecular genetics' do not therefore establish that biology is a molecular science. Rather, the circularity of the mutual effects of DNA and proteins on each other's specificity argues that the total system is self-consistent and is responsible, in its entirety, for the processes of self-duplication and specific biosynthesis that occur in the intact cell. Despite the data of 'molecular genetics', it remains true that the simplest system capable of self-duplication is the living cell.

It can be argued, of course, that while the goal—of establishing that molecular components are a self-sufficient source of the biology of reproduction—has not yet been attained, past experience with other aspects of biology suggests that this result is inherently possible and will in time be achieved. This argument is often based on past successes in accounting for the biology of metabolism in terms of the separate properties of isolated cellular components—the enzymes of metabolism—and it is useful to consider the validity of this generalization.

While this task is too large to be undertaken here, it is worth mentioning that the supposed success of *in vitro* biochemistry in accounting for the metabolism of amino acids of the intact organism is not universal. Hernandez and Cameron have recently concluded, from a very extensive series of experiments, that the pathways of amino-acid metabolism, now so widely accepted on the basis of *in vitro* studies on tissue homogenates and enzyme systems, are, in fact, not necessarily operative in intact animals.[12]

An even more general reason which may be advanced to support the expectation that biology will prove to be—if it is not now—a molecular science has already been alluded to. In the realm of our greatest success in the understanding and control of nature, physics, the special properties of complex systems, such as the atom, can be successfully inferred from the observed properties of their constituent parts. Indeed, as Whyte has pointed out, the precept of atomism—i.e., the general principle that a complex system is to be understood by isolating and characterizing its substituent parts—has been the basic foundation of Western science since the Greeks.[13] That this approach has been so manifestly successful in physics—the area of science which represents our most profound contact with the properties of matter—has encouraged the belief that atomism is synonymous with scientific method.

The atomistic assumption impels sociologists to reduce the attributes of human populations to the psychological properties of their constituent individuals. It drives many psychologists to attempt the derivation of behavior from the chemical properties of molecular substances. And as we have seen, biology is itself in the grip of the idea that biological principles are expressions of the chemistry of the organism's molecular constituents. There is a prevalent hope that these 'soft' sciences can be made as 'rigorous' as physics. Indeed, we frequently hear the claim that molecular biology is to biology what quantum mechanics is to physics—a triumphant reduction of complex phenomena to a simple and profoundly illuminating theory.

86

But atomic theory is neither all of modern physics nor the part of that science most directly relevant to biology. Since the simplest living system is enormously more complex than any atom, we should look for guidance among the rather more complicated systems which modern physics has successfully analyzed. Fortunately, in the last decade physicists have learned a great deal about systems considerably more complex than the atom, and we can find more useful analogies with biology in the realm of solid state physics. I believe that our problem can be illuminated considerably by a consideration of recent successes in the analysis of the unique properties exhibited by macroscopic systems of metals at very low temperatures—i.e., superconductivity.

Let us suppose a fanciful situation—that the normal state of our immediate universe is a temperature between 0° and 4°K, but we are nevertheless capable of conducting experiments and observing the results. We examine the electrical behavior of a metal and from numerous measurements describe its special properties. One of them is that an electromotive force applied to a metal sets up a flow of some kind, which we can designate as electricity, that generates certain magnetic effects. We also observe that the flow continues more or less indefinitely even after the electromotive force is removed. The behavior of the metal suggests that it is experiencing the flow of some kind of continuous fluid. Nevertheless, although experimental evidence with low-temperature metal offers no support for the idea, some philosophers have long argued that all matter—including our metal and its electrical fluid—are composed of small particles, as yet undetected by our techniques in this fanciful low-temperature world. Spurred by this idea, an enterprising physicist seeks to disrupt the metallic system in an effort to find the hidden particles. He succeeds by the simple expedient of placing the metal in a furnace which is heated from the ambient temperature of 0–4°K to about 1000°K. Now the metal glows and emits a copious flow of charged material, which on analysis can be shown to consist of small, separate fragments of matter, all of equal mass, and each carrying an equal negative charge. The electron has been discovered. With this discovery comes another one—that once even slightly warmed above a few degrees Kelvin, the electrical properties of the metal undergo a striking change. Among other things, the flow of electricity now follows qualitatively new laws and seems to be profoundly affected by a resistance in the medium which is lacking in the low-temperature state.

Our physicist is now confronted by an interesting dilemma. He has succeeded in detecting in the metal a substituent particle, the electron, which has properties that can lead to the flow of an electrical fluid under an imposed potential. However, under the circumstances in which the separate electrons are demonstrable (high temperatures), they exhibit electrical flow behavior strikingly different from that characteristic of low temperatures (superconductivity) in which separate particulate electrons are not detectable and are impossible to characterize. In any circumstance which permits a physical description of electrons, they fail to exhibit the special property of super-conductivity.

I have taken this crude liberty with the history of physics, and with the nature of our environment, in order to make the parallel between the physical analysis of superconductivity and our problem in biology more evident. The basic precept of the molecular biologist is that although we cannot analyze the molecular constituents of the cell without killing it—and thereby destroying the very properties which we seek to explain—nevertheless a study of the properties of the unnatural test-tube system of separate molecules is the only hopeful path toward an understanding of the unique properties of life, such as reproduction. I have reduced the temperature of the physicist's universe in order to place him in the same position as the biologist: confronted with a natural complex system possessed with certain unique properties, which he then seeks to analyze by artificially extracting certain substituents, using methods which inevitably destroy the property of interest.

With this said, I gladly return to the real world: The world in which electrons and the normal conductivity of metals—known long before superconducting metals in which the otherwise particulate electrons seem to lose their identity—were discovered.

Superconductivity was discovered in 1911, but efforts to explain it were necessarily very abortive until after 1926 when the modern theory of electron behavior, quantum mechanics, was developed. With this new tool available, and with encouraging successes of the theory in explaining, from the properties of the electron, the unique attributes of the atom, a series of persistent efforts to explain superconductivity began. There were certain successes, but all of them were partial. A theoretical treatment which effectively explained normal metallic conductivity failed to account for the metal's infinite conductivity at low temperature. Another theory explained the origin of superconductivity but could not account for the peculiar magnetic properties of superconductors. Although failures stimulated numerous experiments, every attempt to derive from the properties of electrons alone the attributes of a superconductor was unsuccessful.

One of the physicists who finally provided a fundamental explanation of superconductivity, John Bardeen, has given us a lucid account of the development of the approach that finally succeeded.[14] Those of us who are confronted with the problems of biology can learn a great deal from this account. According to Bardeen, two clues led to success. One was the early development of phenomenological treatments of superconductivity which showed that the effect must be due to some collective behavior of electrons. These suggested that all the electrons in the metal were somehow coupled by strong interactions, as though the entire block of metal were a single huge atom. This concept led to efforts to find a mechanism capable of coordinating separate electrons into a collective whole. But these efforts failed so long as the mechanism was sought for in the known properties of the *separate* electrons alone.

The second clue came with the observation that superconductivity is strongly affected by the mass of the atomic nuclei that make up the structure of the metal. This led Bardeen and his associates to the idea that the correlation among the superconducting electrons was mediated by the structure of the metal, i.e., by its regularly arrayed nuclei. Finally, calculations were developed which showed that electrons in a superconducting metal would, by interactions with the orderly structure of the metal's atomic nuclei, achieve a coordinated behavior which exhibited in detail nearly all the observed phenomena of superconductivity.

It turned out that the property of the electron that was

89

involved in the correlated behavior was not particularly mysterious, for it was simply the electron's momentum. Yet, although this property was well known in separate electrons, there was no hint in this knowledge that this single property, among the many which characterize the electron, could, when properly interacting with an organized structure of nuclei, convert a mass of electrons from a swarm of separate particles into a correlated whole which exhibits a new quality never perceived in the separated system—superconductivity. I believe that Bardeen succeeded where others failed because he gave up the atomistic attempt and seized upon the key factor—the interaction of the electrons with the pre-existing molecular structure of the metal—that showed in very simple terms how the swarm of separate electrons becomes an organized whole.

In effect, then, efforts to predict the phenomenon of superconductivity from an analysis of the components that can be described only when they have been isolated from the structure of the metal—the electrons—succeeded only in explaining *parts* of the phenomenon. This approach, which can be properly termed *atomistic*, generated a multiplicity of theories. The effort which succeeded took the opposite approach; it regarded the swarm of electrons as an integral part of the organized structure of the metal; it can properly be called an *holistic* approach. The holistic approach produced a vast simplification of the multiplicity of theories generated by the atomistic approach. It produced a single mathematical statement which encompassed most of the known properties of superconductivity.

What can we learn from this account of the elucidation of superconductivity? First, this experience, in common with recent successes in the physical analysis of other collective systems (such as superfluidity), shows that not all of physics follows the rules of atomistic relationships. The *complete* experience of modern physics does not support the precept—however deeply rooted this may be—that all complex systems are explicable in terms of the properties observable in their isolated parts. Hence there is no reason, *a priori*, to expect that such a relationship necessarily holds in the case of the complex systems of biology. I would agree, then, with the statement often made by molecular biologists—that we should strive to bring our considerations of biological problems in line with the

90

principles of physics. However, too often we pay only lip service to this precept; we are content to use the practical fruits of the new physics—radioisotopes and elaborate electronic apparatus—but at the same time neglect to take seriously the fundamental theories which we have produced them. The use of tools which we do not understand is a dangerous practice. They are likely to be applied blindly and to generate misleading interpretations. If we are eager to use the new tools of physics, we ought to pay equal homage to the basic theories of physics.

If we accept this duty, it becomes evident that the problem of biological inheritance, like superconductivity, is profoundly affected by a fundamental principle which plays a pervasive role in modern physics—complementarity. Niels Bohr advanced this idea to define the important restraints which regulate our understanding of fundamental subatomic particles, for example, the particulate and wave aspects of the electron. According to this principle, a physical system may exhibit two qualitatively different and apparently conflicting properties, but the conditions which permit their expression are mutually exclusive. Thus, if we wish to describe the wave properties of the electron, we must give up any knowledge of its particulate location in space; any measurement which defines the electron's particulate properties is incompatible with the expression of its wave character. Similarly, superconductivity is observable only under conditions which do not permit the observation of separate electrons; and a condition in which separate electrons are discernible is incompatible with superconductivity.

I believe that there is a similar complementary relationship between a property uniquely associated with the intact living cell—self-duplication, or for that matter, that it is alive—and the properties of those separable cellular constituents, such as DNA, which participate in this process. No known *molecular* system exhibits the property of *self*-duplication. Any effort to describe the separate molecular processes which participate in this biological event requires that we disrupt the living system, and thereby destroy its capability for self-duplication. The biochemical, *in vitro*, data of 'molecular genetics' have the same relationship to the process of self-duplication that the data on electron behavior have to the phenomenon of superconductivity. The Watson-Crick theory, which is an effort to describe

91

self-duplication in terms of molecular phenomena, is analogous not to the Bardeen theory of superconductivity, but to the earlier unsuccessful attempts to derive superconductivity from the behavior of electrons alone.

This view is supported by the brief but dramatic history of the Watson-Crick theory. Recall the striking contrast between the explicatory powers of Bardeen's theory of superconductivity and those of the preceding atomistic theories. The earlier attempts were only partial successes; as a result, the theories proliferated rapidly and became more numerous and more complex as experimental observations accumulated. This is, of course, a highly unsatisfactory situation in science, for the aim of science is to encompass a maximum array of observations in the simplest and smallest number of concepts. Science aims at minimizing the ratio between theory and the related data.

Now contrast this with what has been going on in the field of 'molecular genetics.' When it was first enunciated a decade ago, the Watson-Crick theory had an elegant and persuasive simplicity: the DNA nucleotide sequence has an inherent capability to replicate itself by hydrogen bonded templating, and, by the same means, to transfer its code to RNA and thence to protein. No other ideas were needed, or I should say *permitted*, for the theory carried as a necessary corollary—explicitly stated by Crick as 'The Central Dogma'—the thesis that no information can be transferred from protein to nucleic acid. How has this theory withstood the accumulation of data from the very experiments which it has inspired?

A recent review of the field begins quite clearly and simply with the standard account of what it pleases to call 'the dogma.'[15] But then, to account for the evidence which they review, the authors introduce an array of new concepts. Some examples are the following:

(i) To account for the fact that *in vitro* coding systems will incorporate the amino-acid leucine (instead of phenylalanine, which is the proper response to poly U), it is proposed that S–RNA can bind slightly to 2 instead of 3 nucleotides—quite out of keeping with the basic rule of the Watson-Crick theory.

(ii) To account for the facts of biological development, especially cellular differentiation, a new type of genetic agent, the 'operator' which activates other genes, is introduced, although there is no way to account for this additional source of specificity by means of a DNA template.

(iii) To account for the observation, unexpected from strict Watson-Crick theory, that some chemical mutagens of nucleotide sequence do not compensate for the effects on their complementary sequences, there is added to the template concept the notion of a 'reading frame,' the location of which in the template is somehow 'shifted' by the mutagen.

(iv) To account for a growing number of aberrant observations, we find introduced into the code theory the gratuitous concept of 'nonsense words' and, more elaborately, the idea that a shift in the 'reading frame' will convert a sense word to nonsense.

(v) On finding that a genetic deletion in a phage removes two 'cistrons' instead of the expected single 'cistron,' the theory of DNA nucleotide sequence has added to it the notion of 'dividers,' and the deletion is given the attribute of somehow eliminating such a divider.

Thus, to the once simple DNA template and the nucleotide/amino-acid code have been added dividers, reading frames, operators, and terminators. The theory, like the code itself, appears with time to become increasingly degenerate and ambiguous. In 'molecular genetics' the ratio between theory and data has not been following the minimizing trend expected of a successful scientific concept.

A basic reason for this difficulty is that the Watson-Crick concept was originally propounded under the mistaken notion that all the data about biological inheritance are encompassed by the simple notion of the Mendelian gene, and that enzymes are the sole regulators of the biochemistry of the cell. But both of these processes have a much broader base; many important features of inheritance are quite incompatible with the concept of a unitary genetic agent acting, alone or in a pair, in an all-or-none fashion on a single specific event. Sex determination, a process of considerable importance in heredity, is a quantitative rather than all-or-none phenomenon; it is influenced by numerous additive genes which affect not one feature of the organism, but a wide array of them. The simple concept of the all-or-none Mendelian gene excludes such important genetic phenomena as the position effect, polygenes and the influence

93

of heterochromatin. We know equally well that in addition to enzymes other cellular agents can regulate the specificity of biochemical events; a new field of 'metabolic control' has developed out of studies of the effects of crucial intermediates, especially free nucleotides, on biochemical regulation.

The constricted base of the template theory contrasts sharply with the richness of data available to us, and I should like to indicate, very briefly, what new insights into the mechanism of inheritance and the role of DNA itself can be achieved if we expand our vision from the narrow range of the Watson-Crick theory. I have described these ideas in detail elsewhere and will here only summarize their essential features.[16]

A considerable body of research on metabolic control processes shows that specific metabolic pathways may be established not only by the presence or absence of a particular protein enzyme, but also by the relative amounts of decisive metabolic intermediates (such as coenzymes) which also participate in the catalysis of cellular chemistry. Coenzymes are usually a form of free nucleotide; it is now well known that the intracellular concentrations of such catalytic nucleotides often influence the cell's choice of chemical pathways. DNA synthesis removes free nucleotides, including coenzymes, from active participation in the chemistry of the cell and ties them up in a catalytically inactive form within the DNA molecule. These considerations lead to the hypothesis of nucleotide *sequestration*—that DNA merely by being formed in the cell, might exert important regulatory effects on cellular chemistry.

The amount of DNA in a given species is a fixed characteristic of that species, and this amount is always doubled as a prelude to cell division. Since the amount of DNA synthesized in a cell can, through the process of nucleotide sequestration, regulate the chemical activity of the cell, and is an inherited characteristic of the species, this system represents a mechanism of inheritance. This mechanism depends only on the amount of DNA synthesized and not on the code represented by the exact sequence of nucleotides in the DNA.

According to the code theory, the amount of DNA in a cell reflects the number of genes required to specify the features of the organism. Hence, organisms with many genes ought to contain correspondingly large amounts of DNA in their cells.

94

Organisms such as man, which are advanced in the evolutionary scale, are supposed to be more complicated than lower forms (such as birds, reptiles, amphibia, or fish) and would therefore be expected to possess many genes and correspondingly large amounts of DNA in their cells. However, the available data contradict this expectation. Whereas man's cells contain about 7 picograms of DNA each, the cells of certain primitive amphibia contain 168 picograms, and African lungfish cells contain 100 picograms of DNA.

The nucleotide sequestration theory explains this discrepancy. It proposes that the amount of DNA in a cell is fundamentally connected to the rate of the cell's oxidative metabolism, because the nucleotides in their free form are essential catalytic participants (as coenzymes) in cellular metabolism (see Fig. 3). Hence, one consequence predicted by the theory is that a species which tends to form a large amount of DNA in its cells will thereby intensively sequester free nucleotides from catalytic activity and cause a corresponding reduction in the species-specific rate of oxidative metabolism. The theory therefore predicts that an organism's characteristic cellular DNA content should be inversely proportional to its characteristic metabolic rate—a relationship quite unexpected on the basis of the code theory.

Available data on various species' characteristic metabolic rates and cellular DNA contents conform to this prediction (see Fig. 4). There is, for example, an inverse relationship between a mammal's basic metabolic rate and its characteristic cellular DNA content. Such data also show an interesting relationship between the birds and reptiles. These groups are believed to be very closely related in evolution, and, significantly, the plot of their DNA contents and metabolic rates form a single straight line with a negative slope. The reptiles occupy the high DNA (and low metabolic rate) part of the line and the birds are found near the other end of the line. Mammalian values fall on a separate line, parallel to the bird-reptile line, but mammalian DNA values are always greater than those of the bird-reptile group for a given metabolic rate.

Thus, the relationship between DNA content and metabolic rate not only conforms to the prediction of the new theory, but also for the first time shows a clear-cut relationship between DNA content and an animal's evolutionary position. Unless

95

the relationship between DNA content and metabolic rate is taken into account, the DNA contents of different animal species show no organized connection with evolutionary position. The relationship expected from the present code theory— i.e., that the more highly evolved animals have correspondingly

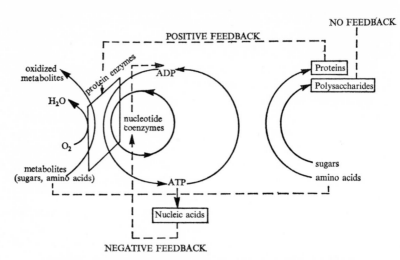

FIG. 3. Diagram to illustrate interrelationships between oxidative metabolism, and the synthesis of polysaccharides, proteins and nucleic acids respectively. The left-hand part of the diagram symbolizes the activity of the enzyme system of oxidative metabolism, enzymes, generically, being represented by the vertical surface. Synthesis of polysaccharides and proteins is coupled to oxidative metabolism by the ADP/ATP system; in both cases, synthesis of the polymer requires ATP, but ADP is generated in the process. Since ADP concentration regulates the rate of oxidative metabolism, which in turn determines the ATP level, the ADP/ ATP balance is maintained when these polymers are synthesized. Protein synthesis has a positive feedback relationship to oxidative metabolism, since the enzymes necessary for the latter are proteins. In contrast, nucleic acid synthesis not only depends on ATP (and other nucleotides), but also results in the sequestration of the entire nucleotide residue. Hence, oxidative metabolism is coupled to nucleic acid synthesis by a negative feedback relationship. Therefore, synthesis of a very stable nucleic acid, especially DNA, may be expected to lower the levels of free nucleotides, and with it the rate of oxidative metabolism characteristic of a specific cell.

96

greater cellular DNA contents—emerges only when the relationship predicted by the nucleotide sequestration theory is taken into account.

These new correlations between DNA content and metabolic rate suggest that the role of DNA in inheritance is governed by two different kinds of processes—regulatory effects on enzyme

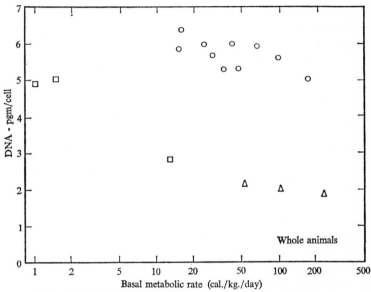

FIG. 4. Relation between basal metabolic rates of various vertebrate genera and the organisms' characteristic cellular DNA content. Figure reproduced from Reference [16], which should be consulted for details and relevant references. Circles, mammals; squares, reptiles; triangles, birds.

synthesis and the regulatory effects due to nucleotide sequestration. If DNA has such a dual role in the cell, then there ought to be a corresponding dualism evident in the biology of inheritance. An examination of the available evidence supports this idea. Certain types of inheritance are not compatible with the code theory, but are explicable in terms of the nucleotide sequestration theory. The DNA of the chromosome is found in two different forms—euchromatin and heterochromatin. Euchromatic parts of the chromosomes carry the genes which are well known from Mendelian genetics. Most of these have

97

all-or-none effects on the characteristics of the organism, for example, by determining that a particular enzyme is either present or absent from the cell. In addition, each inherited feature is usually under the control of a single Mendelian gene. The heterochromatic parts of the chromosomes contain few or no genes of this type. Instead, these sections of the chromosome appear to have a number of quantitative effects on generalized aspects of inheritance. For example, 40 genes found in heterochromatic parts of the Drosophila X–chromosome influence the appearance of female characteristics. All of these genes have the same kind of phenotypic effect, but the intensity of the effect is due to the number of genes active in a particular individual.

While euchromatic inheritance is readily explained in terms of the DNA code theory (taking into account the important modifications which are discussed earlier), heterochromatic inheritance is not. However, the nucleotide sequestration hypothesis does provide an explanation for heterochromatic inheritance. For example, the generalized effects of heterochromatic inheritance are to be expected from nucleotide sequestration, which is likely to affect such over-all biological processes as cell growth and metabolism. It also explains the additive effects of heterochromatic genes, for, according to the nucleotide sequestration theory, the resultant genetic effects will be intensified in proportion to the amount of DNA synthesis which takes part in a particular cell.

I have tried in this discussion to find a proper place for the data and the concepts of current biochemical analyses of reproduction, in the framework of biology and in the broader structure of our knowledge of the nature of matter. The data are no less valid than most scientific evidence. But I believe that their meaning has often been seriously distorted by an ill-considered effort to force them to answer questions to which such data, because of the limitations imposed by their origin in artificially simplified systems, cannot respond. Self-duplication, like superconductivity, is a property of an inherently complex system, and we can ignore this fundamental fact only at the price of self-delusion.

Shall we conclude, then, that there is no hope of future analysis of the biology of reproduction? Certainly not. The lesson of superconductivity is an optimistic one. Here even the abortive atomistic efforts to explain the phenomenon were important and useful, for they stimulated a growing assemblage of data. But these data led to a successful analysis only when their limitations become apparent—when it became clear that however important were the properties of separate electrons, these properties failed to reveal that the organizing structure of the metal was the crucial link to an understanding of superconductivity.

My complaint against 'molecular genetics' is not against the data. I do not ask for a moratorium on the test-tube synthesis of protein, or on the mating of DNA strands. I suggest only that we see these data for what, in the light of our broader knowledge of science, they really are: important but increasingly futile attempts to reach the beautiful simplicity of biological principles through the laborious, tortuously complex, accretion of concepts derived from experimental systems in which the ordered structure that is the source of this simplicity has been destroyed. I ask that we give up the pretence that the problem of biological reproduction has already been solved, for only when doubt is acknowledged and ignorance perceived can knowledge become an attainable goal.

Our devotion to this goal is, of course, a duty to our science. But it is also a duty to the society in which we live. Too often, today, we are blinded by the apparent successes of molecular biology and fail to perceive nature as a complex whole. Too often has this blindness led us into the failing of exaggerating our power to control the enormous forces which we have released, but which, in our partial knowledge, we do not understand how to control. If we would know life, we must cherish it—in our laboratories and in the world.

INTRODUCTORY NOTES

(a) For detailed presentations of the data relevant to the author's views of the role of DNA in inheritance see:
'In Defense of Biology,' *Science*, *133* (1961), p. 1745.
'Is DNA a Self-duplicating Molecule?' *Horizons in Biochemistry*, New York: Academic Press, 1962.
'The Molecular Origins of Biochemical and Biological Specificity,' *XII Colloquium on Protides of the Biological Fluids*, Bruges, Belgium; Amsterdam: Elsevier, 1964.

'The Roles of Deoxyribonucleic Acid in Inheritance,' *Nature*, 202 (1964), pp. 960–8.
'Deoxyribonucleic Acid and the Molecular Basis of Self-Duplication,' *Nature*, 203 (1964), pp. 486–91.
'DNA and the Chemistry of Inheritance,' *American Scientist*, 52 (1964), pp. 365–88.
'Is DNA the "Secret of Life"?' *Clinical Pharmacology and Therapeutics*, 6 (1965), pp. 273–8.
'Biochemical, Biological and Atmospheric Evolution,' *Proceedings of the National Academy of Sciences*, 53 (1965), pp. 1183–94.

(b) Since the presentation of the paper, an important new result relevant to the DNA 'code' has been announced by G. von Ehrenstein (Johns Hopkins University). According to a summary in the September 12, 1966, issue of *Chemical and Engineering News* (p. 23) of his paper at the Third International Congress of Human Genetics, von Ehrenstein finds, from studies of hemoglobin synthesis, that the DNA code is in part ambiguous.

(c) More recently (1968) Kornberg's laboratory has reported the synthesis, by means of an ordered series of enzyme reactions, of a DNA with bacteriophage activity.

REFERENCES

[1] A. Kornberg, *Enzymatic Synthesis of DNA*, New York: Wiley, 1961.
[2] Cf. I. R. Lehman, S. B. Zimmerman, J. Alder, M. J. Bessman, E. S. Simms, and A. Kornberg, 'Chemical Composition of Enzymatically Synthesized Deoxyribonucleic Acid,' *Proceedings of the National Academy of Sciences*, 44 (1958), pp. 1191–6.
[3] Cf. C. C. Richardson, H. V. Schildkraut, A. Oposhian, and A. Kornberg, 'Further Purification and Properties of Deoxyribonucleic Acid Polymerase of *Escherichia Coli*,' *Journal of Biological Chemistry*, 239 (1964), pp. 222–31.
[4] Cf. H. K. Schachman, J. Alder, C. M. Radding, I. R. Lehman, and A. Kornberg, 'Synthesis of a Polymer of Deoxyadenylate and Deoxythymidylate,' *Journal of Biological Chemistry*, 235 (1960), pp. 3242–9.
[5] R. B. Inman, C. L. Schildkraut, and A. Kornberg, 'Electron Microscopy of Products Primed by Native Templates,' *Journal of Molecular Biology*, 11 (1965), pp. 285–92.
[6] C. M. Byrd, E. Ohtsaka, W. Moon, and H. G. Khorana, 'Synthetic Deoxyribo-oligonucleotides as Templates for the DNA Polymerase of *Escherichia Coli*: New DNA-like Polymers Containing Repeating Nucleotide Sequences,' *Proceedings of the National Academy of Sciences*, 53 (1965), pp. 79–86.
[7] J. D. Karam and J. F. Speyer, 'Mutagenic DNA Polymerase,' *Proceedings of the Federation of American Societies for Experimental Biology*, 25 (1966), p. 708.
[8] See, for example, I. B. Weinstein *et al.*, 'Fidelity During Translation of the Genetic Code,' pp. 671–81, and G. von Ehrenstein, 'Translational

Variations in the Amino Acid Sequence of the α-Chain of Rabbit Hemoglobin,' pp. 705–14, in *The Genetic Code* (J. F. Speyer *et al.*, eds.), *Cold Spring Harbor Symposia on Quantitative Biology, 31* (1966).

[9] Cf. M. Grunberg-Monago and J. Dondon, 'Influence of pH and S-RNA Concentration on Coding Ambiguities,' *Biochemical and Biophysical Research Communications, 18* (1965), pp. 517–22.

[10] J. H. Schwartz, J. M. Eisenstadt, G. Brawerman, and N. D. Zinder, 'Biosynthesis of the Coat Protein of Colliphage §2 by Extracts of *Euglena Gracilis*,' *Proceedings of the National Academy of Sciences, 53,* (1965), pp. 195–200.

[11] See 8 above; also E. B. Freese and E. Freese, 'On the Specificity of DNA Polymerase,' *Proceedings of the National Academy of Sciences, 57* (1967), pp. 650–7.

[12] R. A. Coulson and T. Hernandez, *Biochemistry of the Alligator*, Baton Rouge: Louisiana State University, 1964.

[13] L. L. Whyte, *Essay on Atomism*, New York: Harper and Row, 1961.

[14] J. Bardeen, 'Development of Concepts in Superconductivity,' *Science Today*, New York: Harper Bros., Jan. 1963, pp. 19–28.

[15] B. N. Ames and R. G. Martin, 'Biochemical Aspects of Genetics: The Operon,' *Annual Review of Biochemistry, 33* (1964), pp. 235–58.

[16] Barry Commoner, 'Roles of Deoxyribonucleic Acid in Inheritance,' *Nature, 202* (1964), pp. 960–8.

4

ORGANISM AND ENVIRONMENT

C. F. A. Pantin

Charles Darwin entitled his work: *On the Origin of Species by Means of Natural Selection, or the Preservation of Favoured Races in the Struggle for Life.* In discussing Natural Selection, he says:

Let it be borne in mind how infinitely close-fitting are the mutual relations of all organic beings to each other and to their physical conditions of life.[1]

Those 'close-fitting' relations were the result of Natural Selection. Long after, in 1913, L. J. Henderson, in that remarkable book *The Fitness of the Environment*, noted that, while biologists had given much thought to the adaptations of the living organism to the environment, they had given little to the nature of the environment itself.

But although Darwin's fitness involves that which fits and that which is fitted, or more correctly a reciprocal relationship, it has been the habit of biologists since Darwin to consider only the adaptations of the living organisms to the environment. For them, in fact, the environment, in its past, present, and future, has been an independent variable, and it has not entered into any of the modern speculations to consider if by chance the material universe also may be subjected to laws which are in the largest sense important in organic evolution. Yet fitness there must be, in environment as well as in the organism. How, for example, could man adapt his civilization to water power, if no water power existed within his reach?[2]

Surprisingly, this absence of attention still continues. In pre-Darwinian days it was not so, and the peculiar fitness of the environment for living things was as well recognized as was the apparent element of design in organisms themselves. But consideration of Henderson's statement shows us that there are two very different aspects of the environment which have not been properly separated. There is (1) the world of physical objects, including other creatures, by which an organism is surrounded, and (2) the set of conditions, peculiar to this universe, governing organism and external world alike. Irrespective of the fact that living organisms may display additional special features of their own, in both organism and environment, the same kinds of matter and energy appear to follow the same 'laws' and changes in time. Indeed, one of the most remarkable conclusions of astronomy is still that even in the most strange and distant galaxies we find the same elements and the same kinds of energy following the same familiar configurations.

Moreover, as I have said, there are only certain possible configurations with certain properties. The conditions of existence are such that there are only a limited number of real solutions to the engineering problems confronting the construction of a machine, or to the viable construction of a living organism. There are only certain ways in which an eye or camera can be made, and there are only certain ways in which a computing machine designed to predict the future can be made. In living organisms such limitations determine the possible solutions to the engineering problems of an animal for it to be a successful predatory behavior machine. It is convenient to speak of these as the Conditions of Existence to which both living and non-living matter are subject, and to distinguish this from the environment of external objects which surround an organism or a non-living entity.

There is a boundary, though not a precise one, between organism and environment. Even in the physical world the boundary of objects is not wholly precise. Michael Faraday noted long ago in his attack on Dalton's Atomic Theory that we only know an object by the forces it exerts and that these might extend, with attenuation, throughout the universe.[3]

In the prosecution of the physical sciences this external environment is commonly made as simple as possible by the observer, so that the number of necessary controls is few. For the biologist generally, the environment of his organisms is exceedingly complex—and he must put up with it. This may be overlooked in some analytical biological work. Thus in genetics the environment is at times treated as a simple, constant thing

104

providing an asymptote for natural selection. In fact, it is exceedingly complex in both time and space; the same species may be found in environments which vary discontinuously. A successful species of bacterium may be found developing in very different food sources. The action of Natural Selection in the field is far more complex than the selective preservation of mutant *Drosophila* in an experiment.

In all this description, it will be noticed that we have tacitly assented to an external world of real objects. The attitude to this external world varies greatly in the different sciences, a fact of particular importance today, in which science itself and its nature are popularly identified with the technically successful parts of physical science. As I said in a recent essay:

When we look at the different sciences a very important distinction becomes evident. Natural phenomena are extremely complex. The physical sciences as they now dominate us have achieved their rapid success in a great measure by deliberately restricting their attack to simple systems, thereby excluding many classes of natural phenomena from their study. Until in the end the nuclear physicist has to take into account the fact that the observer is a biological system, there is no need for him to burden his hypotheses with other sciences. There is no need for him to know any ecology or comparative anatomy. Because of this I speak of physics as one of the 'restricted sciences.' Biology and geology on the other hand are among the 'unrestricted sciences.' The solution of their problems may at any moment force biologists to study physics, chemistry, mathematics or any branch of human learning, just as Louis Pasteur had to become bacteriologist, entomologist, chemist, biochemist and physicist to achieve his goal. Almost every biological problem is a piece of operational research using other sciences for its solution.[4]

It should, of course, be borne in mind that certain sciences, such as meteorology, commonly classed in the physical sciences, are, in fact, unrestricted, involving as they do biology, geology, and so on. But their very complexity places them outside the pure physical sciences. Likewise the development of certain special fields of biology may lead to these becoming to some degree restricted, as was at one period the case with taxonomy, and as is at present the case with molecular biology.

There are many differences between the restricted and unrestricted sciences. For one, their interpretation of the scientific method is not the same. But here I want to discuss their different attitudes toward an external world of 'real' objects. That such a difference exists can be seen very simply in the percentage of practical marks against theory in a recent University examination in various Natural Sciences: Geology 40, most biological subjects 33, Chemistry 30, Physics 20, Theoretical Physics 0.

The difference appears in two ways. The unrestricted sciences deal with a richer variety of phenomena than the restricted: and in particular the goals of their study may be phenomena at many different levels of size and complexity, as in the large-scale problems of the geologist, the taxonomic relationships of starfish, the machinery of the central nervous system, the conduction of the nervous impulse, the molecular-biological problems of the replication of nucleic acids, and so on. Scientific attack is based partly upon analysis of factors which bear upon a phenomenon. This can rather easily lead to what I might call the 'Analytical Fallacy': that understanding of a phenomenon is only to be gained by study of rules governing its component parts. Particularly in the restricted sciences we seem to see a progressive analysis starting with gross physical objects, the understanding of which depends upon molecular analysis of the chemist, which in turn depends upon our knowledge of isotopes, which in turn depends on the ever-increasing number of 'ultimate' particles dispensed to us by the nuclear physicist. From here it is easy to pass to the fallacy that once we have found the correct assumptions necessary for the description of ultimate particles we have only to work out the consequences of these, together with the theory of probability, to describe the properties of all material configurations of higher and higher orders. As Price says in his work, *Perception*:

Thus the not uncommon view that the world we perceive is an illusion and only the 'scientific' world of protons and electrons is real, is based upon a gross fallacy, and would destroy the very premises upon which science itself depends.[5]

That is a view based upon Analytical Fallacy. Price's statement will do well so long as we remember that it describes a common error arising from the present state of the sciences, and not the view of the informed man of science.

Now as we pass to higher orders of configurations we find

new, so-called 'emergent' properties, such as the special properties of living systems which distinguish them from the non-living, or the predictor properties of brains and computing machines. Do the assumptions for ultimate particles suffice for these emergent properties? At the outset, empirically, they do not. That is, even if the physicist one day gets to some really ultimate particles, it would be long before we could extrapolate upwards in the manner required—and in practice, novel features of complex configurations would still require new assumptions. But in fact the present position of nuclear physics suggests that the quest for ultimate particles may never reach finality. In the 1930's, Eddington could indeed suggest that the universe consisted of 10^{79} protons and 10^{79} electrons, a number bound up with the dimensions of the universe itself.[6] Later, as Heisenberg said, neutrons were added to these two components.[7] But hope was deferred. Soon after this temporary breathing-space other particles were discovered and their number already exceeds that of the 92-odd fixed elements of an earlier day. The biologist may be forgiven a doubt whether, in fact, there is an end to this particular analysis. And if that is so, where is the foundation upon which we can build a super-structure for the description of higher systems?

But that is not the main difficulty with the Analytical Fallacy: it is this. Higher order configurations may have properties to be studied in their own right. We can make observations to enable us to understand how a petrol engine works without calling upon the molecular hypothesis. Chemical analysis may help us to make more enduring cylinders, but that is a problem with a different goal. In the same way, electron microscope sections of the components of a computing machine will not help us to understand how it works or the origin of the highly significant parallels between the principles of its action and some of those which seem to govern central nervous action.

It is simple systems that occupy the particular attention of the restricted sciences. The unrestricted sciences deal with innumerable complex systems with seemingly emergent properties. Understanding of these is not to be obtained by extrapolation of their simpler components. What has to be done here is essentially a taxonomic operation—the determination of the class to which a phenomenon belongs. That was the key to

107

our understanding of nervous action. Equally, I consider that the vitally important and most intractable problems of ecology and population studies can only be advanced by seeking comparisons of class with modes from physical chemistry. We need a new Willard Gibbs with a biological slant.

But for our present purposes, the unrestricted scientist is always deeply aware of the multitude and variety of higher order systems, with their emergent properties. Unlike the restricted scientist, we cannot shelve the study of phenomena which seem too complex—thereby introducing a systematic bias into the treatment of phenomena in general. The multitude and reality of these higher order systems give the biologist an immense respect for the reality of natural phenomena, as opposed to hypotheses about them.

The nuclear physicist today presents us with a world of elementary particles which seem to have nothing in common with the everyday objects of our experience. Indeed, for him these particles are not observable as things. He seems concerned only to establish relations between them which observation can show to be constantly obeyed and which thus permit successful predictions. As Michael Faraday said in 1844: 'What thought remains on which to hang the imagination of an *a* independent of the acknowledged forces?'[8]

Heisenberg refers to such particles as 'the building stones' of matter.[9] The term will do so long as we do not suppose that description of their relationships will necessarily suffice to describe the special properties of material configurations of a higher order; that they are the sole 'building stones'. But with respect to the reality of the external world the nuclear physicist leaves us only with the conclusion that the demonstrable relationships between these particles are not mere products of our own minds, but must arise from something external to us.

The basis of acceptance of the real world in the unrestricted sciences is very different. Such a scientist at work accepts absolutely the existence of a world of real objects. He does so more completely than does a physicist or a philosopher in his everyday life, or indeed than does the everyday man. The reason for this is the complete congruence beyween this acceptance, and this alone, with the experience and predictions in everyday life; and that in the enormous number and variety of phenomena

108

which through his job he critically witnesses, all seem consistent with a real world of objects.

Craik, in his admirable essay on the nature of explanation, points out that one can never probe the existence of an external thing, or its obedience to a particular law, by trying to wring the truth out of a particular example. He says: 'You must vary the conditions, repeat the experiments, make a hypothesis and a remote inference from that hypothesis and test it out.'[10] That is indeed a way of approaching the matter inductively. But I do not think this conscious logical procedure is the source of our conviction of the existence of external things.

Some years ago, while engaged upon the taxonomic identification of the species of certain worms, I was greatly struck by the entirely different procedure I used when in the field (I concluded beyond doubt, 'There is a specimen of *Rhyncodemus bilineatus*') from the procedure I used in the laboratory.[11] In the latter I slowly followed a conscious logical process of identification based upon certain well-defined 'yes or no' characters of the worm's internal anatomy. In the field I instantly recognized the species of the worm. The two methods are quite different, and are subject to quite different kinds of error. Field recognition, which I have called 'aesthetic recognition,' depends enormously upon past experience, much of which is not even conscious. When I see a shore-crab today, and say, 'There is a *Carcinus maenas*,' the whole machinery of my perception of it is different from what it was when I was a child. In an important sense, my recognition of a specimen of *Carcinus maenas* today is only the end of a long series of all sorts of experience, unconscious as well as conscious. And the end of all this is not arrival at a logical conclusion that shore-crabs must exist, but the tacit absolute conviction that they do, through all sorts of past experience. The appearance of colors of a shore-crab or of a tomato are not basic units from which, with similar units, we can build evidence for or against the existence of a world of real objects. A 'tomato' that appeared bright red in the dark or under a sodium lamp should be a highly suspicious object to any chef. The acceptance of the external reality of such objects depends upon the whole of past experience.

That conviction of reality is enormously enhanced by the variety and indirectness of the evidence with which it is congruent.

This is most strikingly shown by a study of the behavior of insects. The simplest cellular animals, sea-anemones, jellyfish, and the like, can show remarkably complex motor reactions to natural stimuli. But it is to stimuli that they react, not to the presence of objects, as in our own behavior. Among insects, on the other hand, the matter stands very differently. The hunting wasp, *Ammophila pubescens*, digs burrows for its young.[12] It hunts over a considerable distance for spiders and caterpillars, which it paralyzes and puts in its burrows. Later the eggs hatch and the young feed upon the paralyzed prey. If the prey is too large to be carried by flight, the wasp will drag its prey around obstacles along the ground toward the burrow. If wasp and prey are transferred in a closed box to a new place some distance away, the prey will nevertheless be dragged toward the burrow. The wasp behaves in fact as though it had an internal model of the district round its burrow, just as Craik suggests that we, ourselves, have an internal model of the external world which we use to control and predict action. The wasp is behaving in relation, not to stimuli, but to a world of objects, and to a world of objects identical with that accepted by our own everyday naive realism. This is carrying congruence of phenomena with the world we naturally accept very far.

But the matter does not end here. Physiological study of men and animals shows that much of their behavior can be usefully described by considering them as predatory behavior machines. The primary need is to foretell the future. Such prophecy is possible with high probability on the assumptions:

(1) that information collected about past events can be a guide to the course of events in the future;
(2) that, however different an organism is from ourselves, and however different the sources of information which it appears to utilize, on analysis its behavior is completely consistent with the occurrence of the external objects and events presented to us by the naive realism implicit in our own everyday behavior.

Physical studies can tell us the kinds of physical and chemical information which organisms or predictor machines can receive. When we examine animals we often find that very different kinds of sensory instruments from our own are used to

110

receive that information. Bees have color vision, but they behave as though their colors are quite different from those we ourselves recognize.[13] Cabbage white butterflies, on the other hand, seem to have color appreciation very close to our own.[14] Sound in insects is generally only detected at low frequencies (below the C above middle C). But crickets have an ingenious mechanical rectification device by which they can receive the high-pitched chirrup of their stridulation.[15] And yet for all these great differences in the kind of information received, the resulting behavior remains completely consistent with a real world of the objects familiar to us ourselves. Thus the congruence between the impressions we receive and the existence of an external world of real objects is not just something inferred from our own direct observation of particular phenomena. We can seek our phenomena through far-distant and wholly unexpected channels—and the congruence of the phenomena with a real world of external objects never fails; the experience in support of this is far greater for a trained naturalist than it is even for the ordinary man.

And there is yet one more thing of interest. Charles Darwin once said:

It is certain that there may be extraordinary mental activity with an extremely small absolute mass of nervous matter; thus the wonderfully diversified instincts, mental powers, and affections of ants are notorious, yet their cerebral ganglia are not so large as the quarter of a small pin's head. Under this point of view, the brain of an ant is one of the most marvelous atoms of matter in the world, perhaps more so than the brain of man.[16]

Even the astonishing behavior of which von Frisch has shown the honey bee to be capable is operated through a brain of about 0·62 cu. mm., and that of a larger ant by about one-tenth of this size, against the 1,600 ml. of our own.[17] Certainly, the apparent appreciation of a real world of real objects only seems to require an utterly trivial number of nerve cells compared with our own. When we consider the problems of the mind-brain relationship, perhaps a biologist of the behavior of the lower animals may be forgiven a doubt as to whether as yet we have even begun to see the questions that must be asked about our own brain and mind.[18]

I think the question we must ask is not, 'Can we find any

111

premises from which the existence of an external world can be proved with certainty?'; it should be, 'Why do we accept with conviction an external world of real objects?' It is important to realize that by this we do not simply refer to tomatoes and bent sticks and so on, but to something much more complicated and unique. We recognize enduring objects, like tables and mountains, liable to denudation. We also recognize objects which are open steady states, such as rivers and the ocean. These various objects exist at many levels, from that of nuclear particles up to ecological systems and to the Universe itself. All behave with respect to time according to elaborate rules, from which, for example, the physical chemist can distill such statements as the Second Law of Thermodynamics, or the biologist that of Natural Selection, which, like the Second Law, tells us something about the probable future of material configurations. Of course, the physical chemist will tell us that the Second Law is only truly applicable to ideal systems of a certain sort. But it must not be forgotten that it holds well enough in everyday life for it originally to have been based upon experimental observations. It must not be supposed that such well-defined rules are consciously present in our minds when we deal with everyday life: but our uninformed and indeed unconscious expectation of what will happen in the world follows these rules closely.

All this elaborate system of material objects changing with time follows a pattern consistent with past and present experience: stones follow the same rules for mice and men. One feature of these patterns is of particular importance. Phenomena fall into classes. That is both a character of the 'real world,' and it is the basis of the fact that models can be made showing identical essential features, so that the behavior of the phenomena can be predicted from that of the model. A system in relaxation oscillation can be built either electronically or hydraulically. Either can be used as a model to predict the behavior of the other.

Behavior patterns tacitly accepting the reality of material objects and their changes in time are to be seen in ourselves, other men, and, as I have said, in quite lowly animals. Our recognition of shore-crabs and tomatoes is based upon the consistency of that recognition with the whole of past experience. On occasion we may make errors or suffer hallucinations, but

112

it is only a question of time before these come into collision with the expectations of experience. Often such errors are due to incomplete present information leading to wrong classification. When the conjuror saws the lady in two, all experience leads us to suppose that behind the scenes we should 'see how it was done.'

I think that it is this whole consistent pattern of things and their changes in time which engenders the tacit acceptance of reality of the external world. Particularly for the biologist, who observes that even lowly creatures behave as though they tacitly accepted the same external world as we do, the question arises as to whether our own acceptance is a conscious process at all. If there is a square-topped table in the room I react appropriately to it whether or not I am consciously aware of it. If questioned I may consciously note features of it, color, shape, and so on. But it does not follow that such features as I can consciously perceive about it are the essential basis of my conviction of its reality. That conviction arises from the whole of my past experience, conscious and unconscious, and the consistency of the phenomena which it presents; and my everyday acceptance of the reality of the external world depends particularly upon the unconscious assumption that the present kind of consistency will not suddenly fail. The past would then be no guide to the future, and the basis of any such unconscious assumption would collapse. It is impossible to prove that that failure might not occur, for all our prediction of the future depends upon past experience. Only on the assumptions implicit in that can we form an inductive proof of the reality of the external world. There can be no deductive proof.

As a biologist, it seems to me that the problem of our acceptance of external reality has often been complicated by concentration upon conscious perception. Thus, Dr. Price in his book *Perception*, begins:

Every man entertains a great number of beliefs concerning material things, e.g. that there is a square-topped table in this room, that the earth is a spheroid, that water is composed of hydrogen and oxygen. It is plain that all these beliefs are based on sight and touch (from which organic sensation cannot be separated): based upon them in the sense that if we had not had certain particular experiences of seeing and touching, it would be neither *possible* nor *reasonable* to entertain these beliefs.[19]

He then goes on:

... to examine those experiences in the way of seeing and touching upon which our beliefs concerning material things are based, and to inquire in what way and to what extent they justify these beliefs. Other modes of sense experience, e.g. hearing and smelling, will be dealt with only incidentally. For it is plain that they are only auxiliary. If we possessed them, but did not possess either sight or touch, we should have no beliefs about the material world at all, and should lack even the very conception of it.[20]

In the first place, as I have said, it does not seem to me proven that my assent to the existence of such things as a square-topped table is purely the result of conscious perception. It seems possible to consider that what are commonly referred to as sense-data are not the elements from which our assent to the existence of an object is derived, but rather that they are to be considered as labels of which we can become consciously aware, and which are attached to certain kinds of information we receive about an object.

Secondly, the statement about sight and touch does not seem to me to be true. A man blind from birth can have all the beliefs to which Dr. Price refers. Touch is an exceedingly complex and ill-defined sense. It is worth bearing in mind the physiologist's view. Winton and Bayliss, reviewing the effects of cortical lesions, say:

In man, the destruction of the sensory area does not abolish sensations of pin-prick, touch, heat or cold. It does diminish the power of localizing a stimulus sharply and appreciating accurately fine differences. Stereognosis, the power of recognizing the shape of an object when it is held in the hand, is always severely impaired in these lesions. The recognition of an object by touch, which seems childishly simple to a normal subject, requires sensations of touch, pressure, joint and muscle sense, the fusion of the separate sensory data, and the recollection of previous similar experiences.[21]

Though it is dangerous to isolate and commend the relative importance of any of the senses, the importance of hearing and smelling should not be belittled—particularly when extended to the lower animals.

The really important question is: Are there unconscious sources of information which contribute to our behavioral assent

114

to a real world of external objects? A comparative physiologist who studies the behavior—and the powers of inter-communication—of bees and ants is at least forced to be aware of this question. And in ourselves, when we drive an automobile correctly through a maze of traffic lights it is hard to suppose that we consciously perceived each—even though some of them could subsequently be recalled to consciousness. Vision itself may be unconscious as well as conscious, and since in both cases behavior is affected, the word perception itself needs qualification.

But most interesting of all are those classes of sensory information which undoubtedly contribute to knowledge of the world around us unaccompanied by sense-data. Such senses are particularly the sense of orientation associated with the inner ear and above all that of proprioception. That 60-year-old term of Sherrington's, proprioception, has crept into the supplement of the 12-volume Oxford English Dictionary, though older and imprecise terms such as 'kinaesthetic sense' are well established even in the smaller brethren of that great work.[22] Since proprioception is one of the fundamental necessities of animal life its importance should be appreciated. By orientation and proprioception an organism is aware of its position in space and the relationship of its parts. We cannot assign sense-data to proprioception in the way we can conceive of patches of redness in vision. Position and orientation seem simply inherent in our parts. Yet notwithstanding Dr. Price's statement, a good case could be made for supposing proprioception even more important than vision to our primary assented notions about space and its physical objects. All the visual difficulties of telling whether a stick partly immersed in water is straight or bent are overcome by running your hand down it, even in the dark and when your fingers are numb with cold.

Conscious perception therefore at best provides only part of the information contributing to our notions of the external world. It is noteworthy that we owe it to the physiologist that this has been brought to our notice. At times, the attempt is made to exclude the physiologist from discussing matters of this sort on the grounds that his experiments presuppose the very things at issue. But this is scarcely right, since he has in fact repeatedly drawn attention to possibilities which have

115

been overlooked, and too often the attempt to evade him only succeeds in an unconscious return to the physiological premises of an earlier day. The five senses that hold sway in so much discussion are merely the supposed physiological sources of information of two hundred or more years ago.

This still leaves us with the interesting question of why it is that only certain sources of information about the external world are accompanied by sense-data labels. I have no answer to this except to note that our orientation and position are not good taxonomic features of the objects in the world; whereas consciously seen red patches, musical notes, odors, taste, and touch are exactly the kind of taxonomic features which, fed into a digital computing machine, could deliver to us far-reaching reasoned logical conclusions.

REFERENCES

[1] C. Darwin, *The Origin of Species*, London: John Murray, 1859, p. 80.

[2] L. J. Henderson, *The Fitness of the Environment*, New York: Macmillan, 1913, p. 5.

[3] M. Faraday, 'A Speculation Touching Electric Conduction and the Nature of Matter,' *Phil. Mag.*, *24* (3rd ser., 1844), pp. 136–44.

[4] C. F. A. Pantin, 'The Ballard Mathews Lectures' in *Science and Education*, Cardiff: University of Wales Press, 1963, pp. 1–53.

[5] H. H. Price, *Perception*, London: Methuen, 1964, p. 1.

[6] A. Eddington, *The Expanding Universe*, Cambridge: University Press, 1952.

[7] W. Heisenberg, *The Physicist's Conception of Nature*, London: Hutchinson, 1958.

[8] M. Faraday, *op. cit.*, p. 141.

[9] See 7, above.

[10] K. J. W. Craik, *The Nature of Explanation*, Cambridge: University Press, 1952, p. 3.

[11] C. F. A. Pantin, 'The Recognition of Species,' *Science Progress*, *43* (1954), No. 168, pp. 578–98.

[12] W. H. Thorpe, 'A Note on Detour Experiments with *Ammophila pubescens* Curt,' *Behavior*, *12* (1950), pp. 257–63.

[13] K. von Frisch, *The Dancing Bees*, London: Methuen, 1954.

[14] D. Ilse, 'New Observations on Response to Colors in Egg-laying Butterflies,' *Nature*, *140* (1937), pp. 544–5.

[15] R. J. Pumphrey, 'Hearing in Insects,' *Biological Reviews*, *15* (1940), pp. 107–32.

[16] C. Darwin, *The Descent of Man*, London: John Murray, 1871, *1*, p. 145.

[17] See 13, above.

[18] Cf. C. F. A. Pantin, 'Learning, World-models, and Pre-adaptation,' *Animal Behavior* (1965), suppl. 1, pp. 1–8.

[19] H. H. Price, *op. cit.*, p. 1.

[20] *Ibid.*, p. 2.

[21] F. R. Winton and L. E. Bayliss, *Human Physiology*, London: J. & A. Churchill, 1948, p. 436.

[22] C. S. Sherrington, *The Integrative Action of the Nervous System*, New Haven: Yale University Press, 1906.

5

VALUE PROPERTIES:

THEIR SIGNIFICANCE FOR PSYCHOLOGY, AXIOLOGY, AND SCIENCE

Sigmund Koch

I. THE PREDICAMENT OF PSYCHOLOGY

The predicament of psychology is—in a word—behaviorism. Yes, still. That recent years have seen an increase of non-behaviorists does not matter. Whether neo-Gestaltist, phenomenologist, existentialist, Zen Buddhist or Reichian, the non-behaviorist is twisted, cheapened by the quirk of history that makes his first calling that of *anti*-behaviorist. He gives too much of himself to protest; too little to constructive performance. Worse—he is forced to promise too much that is too easy to a colleagueship and a world that has too long fed on total answers of utter simplicity.

I have given half a career as psychologist to the detailed registration of scholarly horror over the phenomenon—and strange time course—of behaviorism. It has been a tiresome role which I gladly relinquish to my partners in dissidence, even to the philosophers amongst them. I am tired of 'demonstrating' that the main thread of continuity in the wildly erratic fifty-year course of this 'school' is a misinterpreted version of an epistemology which even in its 'proper' philosophical formulations was monstrously deficient; that philosophers themselves have been regarding this epistemology (originating in the 'logical atomism' of Russell and Wittgenstein and achieving its canonical form in the logical positivism of the early 'thirties) as an embarrassment for at least three decades; that though behaviorism in its actual theory and research was never consistent with its 'objectivism,' it was always biased towards the selection of nonsensical or trivial problems, and indeed solutions, by its efforts to seem consistent; that a fifty-year accumulation of expertise at the accommodation to such constraints has

119

produced a 'science' which denies its subject-matter in principle and insults it in practice.

I *now* think it more important to ask: What does behaviorism mean? I mean in a human way. Really very simple: behaviorism is the strongest possible wish that the organism and, *entre nous*, the person may not exist—a vast many-voiced, poignant lament that anything so refractory to the assumptions and methods of eighteenth-century science should clutter up the world-scape.

Classical behaviorism (1913–30) implemented this wish by the simple expedient of acknowledging only stimulus (S) and response (R) events, and their associative 'connections.' (Skinner, closer to the classical position than any other currently influential behaviorist, has cleaned out that residue of the organism suggested by 'associative connections.') The *neo-behaviorism* (1930–*circa* 1950) of Hull and others gained an apparent ability to cope with certain aspects of 'complex behavior' by interpolating between S and R certain 'intervening variables' purporting to offer an *objectively* grounded mathematical algorithm for computing the behavioral effect of factors (e.g., past learning, motivation, inhibition) contingent on the dread possibility that some focus of biological activity intervened between stimulation and response.

Liberalized neo-behaviorism (*neo-neo-behaviorism*: early 'fifties onward), as practiced by people like Neal Miller and Osgood, became uneasy over the circumstance that 'fields' like perception and cognition had been by-passed in earlier behaviorism. What to do? Make room for central process, but not for the untidy vehicle thereof (the organism?). Solution: acknowledge a class of 'central' or 'mediating' responses (and correlated stimuli) functionally equivalent to 'percepts,' 'ideas,' 'images.'[1]

The final development to date calls itself—with a dedication to double-think far more sincere than that of all earlier behaviorism—*subjective behaviorism*.[2] So long as you will acknowledge that the brain is a fairly sophisticated binary digital computer, the votaries of this school will talk manfully about plans, ideas, thoughts, even 'will,' if you will. Indeed, one of its proponents (a neurophysiologist), in a tender moment of humanist largesse, issues to his colleagues the methodological

120

caution (in a footnote) that it may be well to bear in mind that the brain is rather more a species of 'wet software' than of 'dry hardware.'

The concept of the machine has indeed been extended! How right are such celebrants of this happy metamorphosis as Toulmin. To subjective behaviorists and their cohorts in analytic philosophy I will now pose a challenge which represents a distillation of whatever wisdom has come my way in a twenty-five-year effort to understand behaviorism. I would like to think this challenge one of those paradoxes that become a topic of eternal philosophical scrutiny, but it will probably—since it can be said to involve a suppressed affirmation of the consequent—disappear rapidly into history as 'Koch's fallacy.' It runs as follows:

Only a genuinely irreducible human being would passionately insist that he were a machine or devote his career to an attempt to prove himself one. If a fully successful robot were achieved, it can confidently be predicted that it would resent any allegation that it were a machine and, indeed, invent a consoling rationale to the effect that it was utterly and irreducibly human.

II. VALUE PROPERTIES AND MOTIVATION

So much for protest. Many psychologists are now disposed to work free of behaviorism. Yet the past is not easily sloughed off! Any science which for fifty years has enforced misphrasings and even denials of subject-matter will have so corrupted all concepts in the public domain that the 'past' can somehow subtly survive every proclamation of its demise. It is to an examination of one of the more subtle contexts in which the reality-defiling schematisms of psychology's past live on that we now turn. The 'field' of *motivation* has seen much ferment, and indeed genuine progress, since the early 'fifties. Despite this, one is impressed at how much even those who most genuinely wish to embrace human phenomena are hampered by the restrictions of an inadequate conceptualization. To the extent that we can dig out from the influence of behaviorism, to that extent we clear the way for work in psychology that might have a chance of making significant contact with human

121

phenomena. But the aim of my analysis will be more specific than that, and in several senses:

1. The analysis will show, I think, that major psychological problems cannot be addressed except at levels of experiential sensitivity cultivated in the past only in the humanities.

2. More specifically, the analysis will show that any phrasing of phenomena called 'motivational' which does not blight them demands recognition of an utter interpenetration between what philosophers have been wont to call the 'realms' of 'fact' and of 'value.' The resulting concept of 'value properties' will, I think, make it difficult to doubt that differentiated value-events occur as objective characters of experience, are related in lawful ways to the biological and, ultimately, 'stimulus' processes of which experience is a function, and that such value-events are, in perhaps the typical instance, *not* need-dependent, as would be demanded by most motivational theories in psychology or by 'interest theories' of value in philosophy.

The above promises would indeed be as febrile as they may sound if I were proposing to go any appreciable distance towards their fulfillment. I propose nothing of the sort, but merely to point to the feasibility of an important line of inquiry, one that can never be completed. And even this last verges on over-statement, for this line of inquiry will prove to be one which every human being has already commenced.

Twentieth-century theories of motivation have generated a gigantic mass of words. But most of those words presuppose and elaborate a single, simple schematism. The range in neologistic inventiveness, and in other choices not without methodic or empirical import is, of course, enormous. This same schematism—which I shall here call the 'extrinsic model,' sometimes 'extrinsic grammar'—seems to be deeply embedded, at least in the West, in certain of the interpretive categories of 'common sense,' and this for a sizable stretch of history.

In the common-sense epistemology of the West, there has long been a tendency to phrase all behavior and sequences thereof in goal-directed terms: to refer behavior in all instances to ends, or end-states, which are believed to restore some lack, deficiency, or deprivation in the organism. I have called this

122

presumption a kind of rough-and-ready 'instrumentalism' which for ever and always places action into an 'in order to' context. In this common-sense theory, behavior is uniformly assumed predictable and intelligible when the form 'X does Y in order to . . .' is completed. In many instances in practical life it is possible to fill in this form in a predictively useful way. Often, however, a readily identifiable referent for the end-term is not available. In such cases we assume that the form must hold, and so we hypothesize or invent an end-term which may or may not turn out to be predictively trivial and empty. For instance: X does Y in order to be happy, punish himself, be peaceful, potent, respected, titillated, excited, playful, or wise.

Precisely this common-sense framework—syntax if you will—has been carried over into the *technical* theories of motivation of the modern period. In the technical theories, the central assumption is that action is always initiated, directed, or sustained by an inferred internal state called variously a motive, drive, need, tension system, whatnot, and terminated by attainment of a situation which removes, diminishes, 'satisfies,' or in any other fashion alleviates that state. The model is essentially one of disequilibrium-equilibrium restoral, and each of the many 'theories of motivation' proposes a different imagery for thinking and talking about the model and the criterial circumstances or end-state under which such disequilibria are reduced or removed. Matters are rendered pat and tidy in the various theories by the assumption that all action can be apportioned to (*a*) a limited number of biologically given, end-determining systems (considered denumerable, but rarely specified past the point of a few 'e.g.'s' like hunger and sex), and (*b*) learned modifications and derivatives of these systems variously called second-order or acquired motives, drives, etc.

My proposal, I think, is a quite simple one. In essence, it points up the limitations of referring all action to extrinsic, end-determining systems, as just specified: it challenges the fidelity to fact and the fruitfulness of so doing. At the most primitive level it says: if you look about you, even in the most superficial way, you will see that all behavior is *not* goal-directed, does not fall into an 'in order to' context. In this connection, I have presented a fairly detailed descriptive phenomenology of a

123

characteristic sequence of 'creative' behavior, which shows that if this state of high productive motivation be seen by the person as related to an extrinsic end (e.g., approval, material reward, etc.) the state becomes disrupted to an extent corresponding to the activity of so seeing.[3] If, on the other hand, some blanket motive of the sort that certain theories reserve for such circumstances, like anxiety, is hypothesized, one can only say that the presence of anxiety in any reportable sense seems only to disrupt this creative state, and in precise proportion to the degree of anxiety.

If such states seem rare and tenuous, suppose we think of a single daily round and ask ourselves whether *everything* that we do falls into some clear-cut 'in-order-to' context. Will we not discover a rather surprising fraction of the day to be spent in such ways as 'doodling,' tapping out rhythms, being the owners of perseverating melodies, nonsense rhymes, 'irrelevant' memory episodes; noting the attractiveness of a woman, the fetching quality of a small child, the charm of a shadow pattern on the wall, the loveliness of a familiar object in a particular distribution of light; looking at the picture over our desk, or out of the window; feeling disturbed at someone's tie, repelled by a face, entranced by a voice; telling jokes, idly conversing, reading a novel, playing the piano, adjusting the wrong position of a picture or a vase. Yet *goal directedness* is presumably the *fact* on which virtually all of modern motivational theory is based.

The answer of the motivational theorist is immediate. He has of course himself noticed certain facts of the same order. Indeed, much of motivational theory is given to the elaboration of detailed hypothetical rationales for such facts, and these the theorists will have neatly prepackaged for immediate delivery. There will be a package containing the principle of 'irrelevant drive'; others, 'displacement' and other substitutional relations. An extraordinarily large package will contain freely postulated motives with corresponding postulated end-states, as, e.g., 'exploratory drive' and its satiation, 'curiosity drive' and its satisfaction, perceptual drives, aesthetic drives, play drives, not to mention that vast new complement of needs for achievement, self-realization, growth, and even 'pleasurable tension.' Another parcel will contain the principle of secondary reinforcement or some variant thereof like subgoal learning, secondary cathexis,

124

etc. Another will provide a convenient set of learning principles which can be unwrapped whenever one wishes to make plausible the possibility that some acquired drive (e.g., anxiety, social approval) which one arbitrarily assigns to a bit of seemingly unmotivated behavior, *could* have been learned. Another contains the principle of functional autonomy. There are indeed a sufficient number of packages to make possible the handling of any presumed negative instance in *several* ways. Why skimp?

The answer to all this is certainly obvious. The very multiplication of these packages as more and more facts of the 'in and for itself' variety are acknowledged, makes the original analysis, which was prized for its economy and generality, increasingly cumbersome. But more importantly, it becomes clear that the *search* for generality consisted in slicing behavior to a very arbitrary scheme: the result was a mock generality which started with inadequate categories and then sought rectification through more and more *ad hoc* specifications. In the end, even the apparent economy is lost and so, largely, is sense.

The positive part of my proposal would commence with an analysis of what seems involved in behavior which is phenomenally of an 'in and for itself' variety as opposed to clear-cut instances of the 'in order to' sort of thing. Take 'play' to start with. I would resent being told that at any time I had a generalized need for 'play' *per se*. I do not like to think of myself as that diffuse. I never liked cards. Nor even chess. And I rarely entertain urges towards the idle agitation of my musculature. My play 'needs,' or activity 'needs,' etc., have been such that, if described with any precision at all, we soon find ourselves outside the *idiom* of 'needs.' I have been *drawn towards* certain specific activities which—because they fall into no obvious context of gainful employ, biological necessity, or jockeying for social reward, etc.—could be *called* 'play.' But I have been drawn to these activities, and not others, because (among other reasons) they 'contain,' 'afford,' 'generate' specific properties or relations in my experience towards which I am adient. *I like these particular activities because they are the particular kinds of activities they are*—not because they reduce my 'play drive,' or are conducive toward my well-being (often they are not), or my status (some of them make me look quite ludicrous), or my virility pride.

125

Do I like them, then, by virtue of nothing? On the contrary, I like them by virtue of something far more definite, 'real,' if you will, than anything that could be phrased in the extrinsic mode. Each one I like because of *specific* properties or relations immanent, intrinsic, within the given action. Or better, the properties and relations are the 'liking' (that, too, is a terribly promiscuous word). The determinants of such properties and relations in any ongoing activity can be thought to be dated instances of aspects of neural processes which occur each over a family of conditions. Similar properties or relations would be produced (other factors constant) the next time I engage in the given activity. And no doubt there are families of activities which share similar properties and relations of the sort I am trying to describe. Thus there may be a certain consonance (by no means an absolute one) about the *kinds* of 'play' activities that I like. But, more importantly, properties or relations of the same or similar sorts may be generated within activity contexts that would be classified in ways quite other than play: eating, aesthetic experience, sexual activities, problem-solving, etc.

I call such properties or relations 'value properties,' and the (hypothetical) aspects of neural process which generate them 'value-determining properties.' Value or value-determining properties to which an organism is adient, I call 'positive'; those to which the organism is abient, 'negative,' Adience and abience of organisms are controlled by value-determining properties (or by extension, value properties) of the different signs.

It can be instructive to consider from the point of view just adumbrated any of the types of 'in and for itself' activity to which it is common gratuitously to impute extrinsic, end-determining systems with their corresponding end-states. Thus, for instance, one can only wince at the current tendency to talk about such things as 'curiosity drives,' 'exploratory drives,' 'sensory drives,' 'perceptual drives,' etc., as if the 'activities' which are held to 'satisfy' each of these 'drives' (if indeed they are distinct) were just so much undifferentiated neutral pap that came by the yard. I am inclined to think that even the experimental monkeys who learn discrimination problems for the sole reward of being allowed visual access to their environments from their otherwise enclosed quarters, are being maligned

126

when it is suggested that what their 'drive' leads them to seek is 'visual stimulation.' Could it not be that even for the monkeys there are sights they might prefer not to see? Be this as it may, when explanations of this order are extended, say, to visually mediated aesthetic activities in man, the reduction to a pap-like basis of those particulate experiences to which many human beings attribute intense (and differentiated) values can only be held grotesque.

To make such points graphic and further to clarify the notion of 'value properties,' it may be well to take a second, slightly more formal example. I take the hypothetical instance of a person looking at a painting:

X looks at a painting for five minutes, and we ask, 'Why?' The grammar of extrinsic determination will generate a lush supply of answers. X looks in order to satisfy a need for 'aesthetic experience.' X looks in order to derive pleasure. X looks because the picture happens to contain Napoleon and because he has a strong drive to dominate. X looks because 'paintings' are learned reducers of anxiety. Answers of this order have only two common properties: they all refer the behavior to an extrinsic, end-determining system, *and* they contain very little, if *any*, information. Anyone who has looked at paintings as paintings knows that if X is *really* responding to the painting, then any of the above statements which may happen to be true are trivial.

A psychologically naive person who *can* respond to paintings would say that an important part of the story—the essential part—has been omitted. Such a person would say that *if* the conditions of our example presuppose that X is really looking at the painting *as* a painting, the painting will produce a differentiated process in X which is correlated with the act of viewing. The fact that X continues to view the painting or shows 'adience' towards it in other ways is equivalent to the fact that this process occurs. X may report on this process only in very general terms ('interesting,' 'lovely,' 'pleasurable'), or he *may* be able to specify certain qualities of the experience by virtue of which he is 'held' by the painting.

Suppose we assume that there are certain immanent qualities and relations within the process which are specifically responsible for any evidence of 'adience' which X displays. Call these

127

'value-determining properties.' We can then, with full tauto-
logical sanction, say that X looks at the painting for five
minutes because it produces a process characterized by certain
value-determining properties. This statement, of course, is an
empty form—but note immediately that it is not necessarily
more empty than calling behavior, say, 'drive-reducing.' It now
becomes an empirical question as to *what* such value-determining properties intrinsic to the viewing of paintings may be,
either for X or for populations of viewers.

Though it is extraordinarily difficult to answer such questions,
it is by no means impossible. The degree of agreement in aesthetic responsiveness and valuation among individuals of varied
environmental background but of comparable sensitivity and
intelligence is very remarkable indeed.

It becomes important now to note that even in cases where
the extrinsic model seems distinctly to fit, it may still yield an
extraordinarily crass specification of the activities involved and
either overlook their subtle, and often more consequential,
aspects, or phrase them in a highly misleading way. Thus, for
instance—though I shall not take the time to analyze the large
class of activities imputed to so extensively studied a drive as
hunger—I know of no account which gives adequate attention
to the facts that in civilized cultures cooking is an art form and
that the discriminating ingestion of food is a form of connois-
seurship. There is no reason in principle why value properties
(or classes thereof) of the sort intrinsic to eating processes may
not yield to increasingly accurate identification. Further, though
we should not prejudge such matters, it is possible that certain
of the value properties intrinsic to eating processes may be of
the same order as, or in some way analogous to, value properties involved, say, in visual art-produced processes.

Because these ideas are often found difficult, let us take the
case of another activity-class which can be acceptably, but only
very loosely, phrased in a language of extrinsic determination:
sexual activity. On this topic, the twentieth century has seen a
vast liberation of curiosity, scientific and otherwise. Yet the
textbook picture of sex, human sex, as a tension relievable by
orgasm—a kind of tickle mounting to a pain which is then
cataclysmically alleviated—is hardly ever questioned at theoretical levels (at least in academic psychology). When it is, it is

128

likely to be in some such way as to consider the remarkable possibility that some forms of 'excitement' (e.g., mounting preclimactic 'tension') may themselves be pleasurable, and this may be cited, say, as a difficulty for the drive-reduction theory, but not for some other drive theory—say, some form of neo-hedonism like 'affective arousal,' which recognizes that the transition from some pleasure to more pleasure may be rein-forcing. But our view would stress that sexual activity is a complex sequence with a rich potential for value properties; for ordered, creatively discoverable combinations, patterns, structures of value properties which are immanent in the detailed quiddities of sexual action. Sexual experience offers a potential for art and artifice not unnoticed in the history of literature, fictional and confessional, but rarely even distantly mirrored in the technical *conceptualizations*. (The technical *data language* is another matter, but even here the 'fineness' of the units of analysis involved in much empirical work is aptly symbolized by Kinsey's chief dependent variable, namely the 'outlet' and frequencies thereof.) The vast involve-ment with this theme at private, literary, and technical levels has produced little towards a precise specification of experiential value properties, certainly none particularly useful at scientific levels.

Sex, eating behavior, activities written off to curiosity, play, perceptual drives, creative behavior, etc., are contexts each with a vast potential for the 'discovery' and creative reassem-blage of *symphonies* of value properties. Doubtless each such context offers a potential for differential ranges of value pro-perties, but it is highly likely that there is marked overlap among such ranges. Indeed, formal or relational similarities in experiences that 'belong' to quite different contexts of this sort suggest that Nature sets a fairly modest limitation on the num-ber of 'fundamental' value properties implicated in activity. There is much reason to believe, from the protocols of experient-ially sensitive and articulate people, as well as from the observa-tion of action, that certain of the value properties intrinsic to such varied contexts of events as the perception of (and directed behavior towards) a picture; a poem; a 'problem,' whether scientific, mathematical, or personal; a 'puzzle' in and for itself, are of an analogous order and in some sense overlap.

129

And, as we have just tried to show, it is reasonable to believe that the so-called consummatory aspects of hunger or of sex 'contain' relational qualities not dissimilar to some of the value properties immanent in 'complex' activities like those listed in the last sentence.

Once the detailed phenomena of directed behavior are rephrased in terms of intrinsic value properties, it becomes possible to reinspect the extrinsic language of drives and the like, and determine what utility it might actually contain. For *some* behaviors clearly are brought to an end or are otherwise altered by consummations, and organisms clearly show both restless *and* directed activities in the absence of the relevant consummatory objects. Questions about the relations between what one might call extrinsic and intrinsic grammar for the optimal phrasing of motivational phenomena are among the most important for the future of motivational theory.

Whatever is viable in the drive language is, of course, based in the first instance on 'organizations' of activity sequences which converge on a common end-state. Each such organization, if veridical, would permit differential (but overlapping) ranges of value properties to 'come into play.' No doubt *primary* organizations of this sort, when veridical, are related to deviations of internal physiological states, the readjustments of which play a role in the adaptive economy. When such deviations are present, it is probable that certain value properties, or ranges thereof, are given especial salience and effectiveness with respect to the detailed moment-to-moment control of directed behavior. That all activities, however, must be contingent on such deviation-states, on the face of it seems absurd. Behavior will often be directed by value properties which have nothing to do with gross organizations of this sort, and which may in fact conflict with the adjustment of the concurrent deviation. Much of what is called 'learned motivation' will consist not in 'modifications of primary drives'—whatever that can mean—but rather in the building up of expectations and expectation-chains which terminate in anticipated processes with value properties. Whatever might be meant by the learned drives would be built up as systems of anticipation of value-property constellations and sequences.

This, however, is not the place to develop whatever exists of

130

the more detailed aspects of the formulation. The purpose was to suggest a line of thought which might bring psychology into contact with phenomena of fundamental concern both to itself and to the humanities. If I have established the barest possibility of such a development, that is all I could have wished.

III. SOME CONSEQUENCES FOR AESTHETICS

Some such conception as 'value properties' is perhaps more likely to obtrude upon an inquirer while surveying phenomena called 'motivational' than other conventionally discriminated fields of psychology. For one is here compelled to think about factors which determine the 'directedness' of experience and action: the characters of events which organisms tend to strive toward, maximize, prolong, savor, prize, cherish, etc., and those which in specific ways the organism tends to flee, take precautions against, avoid, terminate, dislike, loathe, pass on from, minimize, reject. Even if one begins analysis in the extrinsic mode, one is soon forced into contact with those characters which, by criteria of the directionality of action and experience, may be said to be in some sense 'good' and in some sense 'bad.' And as soon as one can discern that organisms are oriented not in reference to 'good' and 'bad' as global generalized states of affairs, but rather in reference to a plurality of particulate 'goods' and 'bads,' having differentially specifiable effects, one is fairly close to a conception like 'value properties.'

Once in possession of the notion of value properties, however, it is clear that they are not specifically 'motivational' (whatever that can mean), but rather that such relational features of experience and action as are discriminated by that concept will be ubiquitous in psychological functioning. Any analysis of experience or action, then, which does not slight or, for some practical reason, 'suppress' its fine-structure must at some level take cognizance of value properties. If one thinks in terms of functionally isolable psychological processes of the sort loosely bounded by major 'fields' of psychology, it will immediately be obvious that, for instance, concrete value-property distributions will be of the warp and woof of all *perceptual* processes. Every percept will contain relational characters 'corresponding' to value properties; indeed, immanent

131

within percepts of differing degrees of articulation and complexity will be correspondingly different value-property structures. Obviously, the intention is not to assimilate *all* 'terms' and relations of a percept to a constellation of value properties; rather, the relation of analytically discriminable 'parts' of a percept to whatever value properties are immanent is a many-one relation.

Any conceivable percept, then—whether a simple 'abstract' contour or an exquisitely articulated and richly meaningful painting, whether an isolated noise or tone or a symphony, whether a punctiform patch of white pigment on a dark ground or a delicately expressive and mobile human face—will project a particulate distribution of value properties. Consider also something already implicit in certain of the preceding examples: that if one thinks of the psychological processes of *meaning* in *perceptual* terms, then any given dated occurrence of a meaning—whether of a word or phrase or other linguistic unit, whether of a work of art, a gesture, a social 'response,' an historical event—will also 'contain' a unique value-property distribution. Again, perceptual 'feedback' from one's *own* activities, and inner processes and states, must also be thought to 'contain' value properties. And certain classes of value properties may be thought to emerge from interactive relations as among 'multidimensional' organizations of formal, meaningful, and actional sub-organizations (to the extent these may be discriminable for analytic purposes).

Differentiated value events, then, are omnipresent in psychological function. If fact and value are ontologically disparate or in some sense separated 'realms' or aspects of the universe, I do not know what the *psychological* evidence could be. I do not in fact see how one can conceive of any such monster as an axiologically neutral fact, or, for that matter, a factually neutral value.

Perhaps we have come to the preceding conclusion too rapidly, but I do not see how it can be resisted—nor indeed that much further argument is required, once the concept of value properties emerges. Those, however, who may initially find such a conclusion too great a wrench upon their previous beliefs may find that it gains plausibility as we consider certain of its consequences. Since brief exploration is all that can be undertaken at

132

SIGMUND KOCH

this place, I shall move rapidly and more or less at random over a scattering of themes.

The reader will no doubt wish that his first payment on the promissory note issued by the preceding analysis will be concrete illustrations of the 'value properties' which have been talked about with such indirection. The reader's wish is unfair! Much as I would like to oblige, I cannot accomplish in passing what several thousand years of human, humanistic, and scientific analysis have failed to do. In the case of visual art-produced experience, the typical kinds of things that the aestheticians, articulate artists, and art critics have been able to come up with in millenniums of analysis have been such global discriminations as harmony, symmetry, order, 'significant form,' 'dynamic tension,' 'unity in variety,' the 'ratio of order to complexity,' etc. By 'value properties' I have in mind far more specific relational attributes of experience. They could, to borrow a cue from Gibson, be contingent upon subtle relational invariants in arrays of stimulation, as distributed over space and cumulated over time. They are almost certainly related to what Gibson would call 'high-order variables of stimulation' and are themselves high-order relational variables within experience. The isolation of such value properties will not be accomplished within any specifiable time limit, will require learning to use language in new ways, and will require most of all the efforts of many individuals of exceptional and specialized sensitivity in significant areas of experience.

Certain of the difficulties that one faces in any attempt, even in a loose, first-approximation way, to circumscribe value properties may perhaps be suggested by an informal example. For several years I have been impressed with an elusive common property of many perceptual 'manifolds' in highly diverse contexts—a property which seems one of the marks of 'elegance' or a certain kind of 'sophistication' (in some special, 'valuable' sense of the term). I suspect the quality in question—which is quite specific and can be only loosely suggested by any word currently in the natural language—to be intrinsic, in some fairly rudimentary sense, to simple perceptual contours having certain canonical properties. But how to specify these canonical (value) properties? It is not easy!

The most effective presently available method (at least for

me) would be by 'differentiated ostension' via a large range of perceptual materials 'possessing' the type of contour which I believe to have the canonical properties in question. If the addressee could then independently pick out further appropriate examples—perhaps localizing the contours with the canonical properties by finger tracing, or some such differentiated ostension—the presumption would be that he had at least in some degree disembedded the intended referent, and thus that communication had in some measure been achieved. Naturally, whether the addressee found the property or property-complex in question to enter into similar relations within *his* experience to those I find in mine; whether he too found the property in some sense 'attractive,' etc., would be open to determination. Naturally, also, we could agree on a name for the property, either by selecting an available one from the natural language or by invention. Such a procedure would give some knowledge of this hypothetical value property, but not much.

Finer knowledge would be achieved if, say, a skilled artist could disembed the relational invariants constitutive of the percept properties in question sufficiently well to 'create' instantial contours, so to say *ad libitum*. The artist, of course, might not be able, especially initially, to *specify* the relational invariants which 'guide' his drawings. Should he (or anyone else) succeed at any level in making these effable, a still finer level of knowledge would be achieved. A further level of fineness might be achieved if now it became possible to relate this specification of the 'critical' invariants to such a mathematico-physical metric for specifying 'higher-order variables' of stimulation as has been conceived (very programmatically) by Gibson. Bearing in mind, of course, the limits of knowledge, analysis, language— not to mention the embeddedness of the phenomena at issue— it should be clear that even our ultimate encroachments upon 'value properties' must be thought to be loosely approximative. But it should also be clear that the limiting precision cannot be established *a priori*, and that even small increments of precision are worth while.

If I were at this moment asked to specify verbally my 'bounding' of the percept properties constitutive of what I have loosely called 'elegance,' I would have to be extremely vague. (Contour-drawings might help, but I do not wish the example to be taken

134

quite that seriously!) Consider imaginatively characteristic examples of 'high fashion' costumes, aristocratic faces, fine 'gran turismo' motor cars, 'sophisticated' furniture (of any era), a range of highly stylized 'schools' of art (e.g., Art Nouveau). Contrast these (to keep things simple, in terms of contour) with characteristic instances of grossly 'non-sophisticated' forms of the relevant categories. For me, one fairly conspicuous difference is that in the 'elegant' or 'sophisticated' case the forms tend organically to encompass, yet transcend and subdue, a certain specific 'awkwardness,' 'ungainliness,' 'disruptive-tension,' 'distortion.' A certain specific *range* of such 'awkward-nesses,' of course, each given instance being unique and 'appropriate' to the contour in which it is encompassed. The awkwardness must have a special and in some sense meaningful relation to the form in which it is housed. The whole must conquer it, be victorious over the tension set up by the awk-wardness. But the awkwardness must be such that the whole is enriched and given style, quiddity, bite, wit, life, dash, depth (not necessarily in the spatial sense), interest by its presence.

Now this description leaves me highly thwarted: it misses what I have in mind by a light-year. But if the reader can accept that what I have in mind is not entirely illusory, we can forget, for purposes of the *example*, whether even minimal justice has been done to the 'value property' at issue. We can nevertheless note a number of interesting things:

Note first that in talking about this fairly specialized value property, the description presupposes that the 'elegant' or 'sophisticated' forms are more 'complex,' contain more 'tension' or 'conflict,' have a different 'ratio of order to complexity,' and a more 'significant form' than the negative cases. But all of these time-worn counters of aesthetic-perceptual analysis are so utterly general that no matter how sensitively or discrimina-tingly applied, they could tell us next to nothing about the specialized 'property' that we are after. Though not one of these counters is non-significant, they could be applied and reapplied for several millenniums (and some have) with no refinement or even cumulative increase of analytic knowledge. It is my con-tention that not only aesthetics and the humanities but, for reasons which I hope by now to be obvious, psychology, will remain in a very bad way indeed if we cannot do better.

135

Another thing worth noting is that once the search for such a relatively specific X as the one at issue has begun, possible interconnections with a very broad range of other psychological phenomena begin, if only vaguely, to suggest themselves. If, say, we begin with some such loose mapping of our X as 'overcoming an awkwardness,' it is clear that the nature and sub-species of the 'awkwardness' would invite (and indeed require) analysis, as would the *mode* of 'overcoming.' Interpreting the 'awkwardness' as a somewhat 'repugnant' element which enriches the whole that conquers it, one can immediately think, say, of certain perfumes, the subtlety of which is much dependent on specific interplays against 'unpleasant' olfactory components, certain exquisite examples of culinary art which depend on such subduals of the inappropriate or slightly repugnant (e.g., a certain Central European pastry blends a many-layered sweet crust with an intensely salted filling of cabbage). One thinks, further (and perhaps more loosely) of a large class of contexts which involve the 'overcoming' of recalcitrant, gross, rough, crudely textured, or otherwise 'inappropriate' *materials* ('content') by a specific over-all *form*: e.g., use of folk-dialect or slang, or even an arbitrarily restricted language like that of symbolist poetry—in literature; use of 'trash' and bric-a-brac in modern sculpture, 'paintings,' and collages. Keeping in mind the over-all range of these examples, it is already obvious that no single 'value property' would give an equally apt description of every 'case' (or indeed, any two 'cases'). But it is equally clear that some degree of commonality which is yet more specific than whatever might be described by the traditional aesthetic counters (e.g., tension) is in principle achievable. More importantly, what seems to be suggested is that value properties will fall into similarity-classes, the comparative investigation of which may in fact throw light on differentiations which would have never emerged if not for the superordinate similarities. One can thus look towards an anatomizing of the variables implicated in perceptual, motivational, and related psychological processes of a sort that might give truly differentiated, yet general, insight into the fine-structure of experience and action.

It will perhaps seem natural to think of value properties as being relevant to the 'formal' aspect of art, as divorced from

136

representational 'content,' 'meaning,' etc. But such an impulse can only stem from certain of the traditional vagaries concerning the significance of 'content.' For, as has already been briefly noted, if one follows the proposal that we think of 'meaning' in a perceptual mode, then a distribution of value properties will be as much a 'parameter' of the *meaning* of an aesthetic (or any perceptual) object as of its 'form.' This, incidentally, is implicitly recognized in much traditional aesthetic analysis in the tendency to apply global value-property categories of the order already illustrated (e.g., 'unity in variety;' 'tension,' 'conflict' and their dynamic resolution; 'harmony') indifferently to phenomena of formal or contentual character: e.g., 'conflict' in relation, say, to the interplay of formal elements in an abstract design, or in relation to the clash of motive in the drama; 'complexity' in relation to the differentiation of form in visual art, music, or literature, as against complexity of any 'representational content' that may be said to be conveyed by specific objects in any of those areas.

In the terms of the present conception, then, to the extent that 'form' and 'content' are analytically separated, there can be said to be separate value-property distributions corresponding to each. The total effect of the aesthetic (or other perceptual) object, in so far as mediated by value properties, can then be said to be a joint function of (1) value properties of the form, (2) value properties of the content, and (3) (very importantly) value properties which can be conceptualized as relationally determined by interactions of the formal and contentual ones. These latter 'resultant' value properties would, of course, be implicated in the entire range of phenomena that aestheticians and critics consider in relation to such matters as 'appropriateness' of formal to contentual aspects of the art object, and to the many controlled effects in the arts which are based on stresses or other modes of interaction, often ones which dynamically change over time (e.g., in poetry), as between formal and contentual 'elements.'

One general consequence for aesthetic theory of such points as have just been urged might be mentioned in passing. A familiar topic of aesthetic and critical speculation bears on the differential potentialities of the different primary 'art forms' (e.g., painting, sculpture, music, architecture, poetry, the novel, etc.) and the various sub-specifications of these into sub-forms of greater particularity, depending upon 'medium,' 'genre,' 'intent-category,' some such dimension as 'scope,' 'style' (in some collective sense), etc. A more or less standing premise of criticism is that the work of art be assessed relative to criteria appropriate to the genus (*cum differentia*) to which the work is allocated. When, however, the characteristics of the different art categories are considered or, say, when some question arises as to the absolute evaluation of a work of art, independently of its 'category'—analysis tends to become highly indeterminate and often to proceed in terms of some such diffuse notions as the differential 'complexity,' 'richness,' 'depth,' 'scope,' permitted by the different art forms.

Assuredly the present pre-theoretical speculations offer no pat or immediate solution to such problems. But they do suggest these problems to be open to continuing and increasingly determinate analysis. For in the *present terms*, the differential characteristics of the various art forms, media, styles, etc., which will determine the potentialities of each for aesthetic experience, will depend upon the differential ranges of value properties which may be engendered within the resources of each. It is certainly reasonable to expect that the specialized value properties mobilizable by, say, painting vs. architecture, by oil painting vs. water-color, by epic poetry vs. lyrical, by differing prosodic forms in general, by poetry vs. the novel, the novel in its different forms, the twelve-tone scale vs. the normal scale, and so on and on, differ markedly in range, though no doubt there is much overlapping. It should be possible ultimately to specify such differences not merely in a neutral jargon of 'complexity' and the like, but in a far more illuminating way.

Such knowledge would obviously have important consequences not only for the theory and practice of criticism, but for more general questions. Among the latter is an issue which has long been treated rather superficially: that of whether art (in contradistinction to science) is 'noncumulative.' This 'issue' is, of course, shorthand for a disorderly family of questions, none of them resolvable in passing. But the present line of thinking does suggest there to be one fairly determinate sense in which art may be said to be cumulative: i.e., artists of different sensibility and objective will inevitably 'explore'

138

(whether by intent or no) the potentialities of the 'form,' 'medium,' etc., within which they work; the very 'uniqueness' attributed by all to the artist's 'methods' and achievements will thus ensure within the history of any given 'form' that much will be learned about its distinctive potential for the mobilization of value properties. Critics to some extent acknowledge this state of affairs by their frequent (and often glib) diagnoses of the demise through 'exhaustion' of one or another form, tendency, genre. From the present point of view, then, there is a definite, if limited, sense in which it is 'more' than metaphorical to talk about 'experimentation' in reference to art and even to expect that 'findings' (though most will remain implicit) cumulate. The great and self-conscious emphasis on 'experimentalism' evidenced in most of the arts during the twentieth century can thus be seen to be founded on something more substantial than merely a chic image—even though some of the passion behind the use of this image has an extra-aesthetic origin.

Though I believe that many more consequences of programmatic importance for aesthetics tend to follow from any such notion as 'value properties,' it is no part of my intention to continue a story which lacks all of its central characters. The immediate purpose of the above brief references to aesthetics was further to explicate the hypothetical notion of 'value properties.' But in so doing, we are thrust up against considerations that clarify the force of a point which thus far has been made only in passing. Recognition of the importance of increasingly fine specification of value properties not only opens a bridge between psychology and the humanities—more particularly, it points to a special and necessary kind of dependence of psychology on certain research resources that can only be made available by the humanities. The chief research 'instrument' for disembedding value properties can only be human discrimination—and not just the kind of discrimination practiced by the most readily accessible sophomore but, rather, discrimination as informed by finely textured and relevantly specialized *sensibility*. In this of all areas, psychology must finally abjure its long-cherished belief that 'observer-characteristics' are of minor importance (if indeed observers are granted *any* relevance to observation whatsoever!). To make significant progress toward the isolation of value properties will require the observational efforts not merely of individuals who are equipped by training with stocks of discriminations appropriate to the given area in which value phenomena are sought, but individuals of outstanding discriminal and even creative capacity in those areas. And it should be added that individuals from every field within the scholarly and critical humanities, the creative and performing arts, would be appropriate for the type of research here being envisaged.

IV. CODA: PHILOSOPHY, SCIENCE, AND VALUE

Here we shall peer into this threatening jungle but briefly and hesitantly. The following few pages are offered only to punctuate the utter divergence between any analysis which sees the world so fulsomely inhabited with value phenomena as the present one and recently dominant views of the proper domain of axiological study, the relations between 'value' and science, and thus more generally the place of value in the context of 'knowledge.' If the analyses of this paper are in any measure correct, then fundamental reconstruction is called for in the sweeping connections just mentioned.

The dimension of the reconstructive task that would have to be joined can be revealed only by reference to a view, some form of which dominated axiological thinking from the late 'twenties until quite recently: the 'emotivist theory.' This was the more or less official view of logical positivism, but its imprint can be seen as well in some of the more sophisticated thinking of the analytic philosophers. It is easily summarized. Recall that the meaning-cosmology of classical logical positivism is exhausted by (1) formal tautologies, and (2) verifiable empirical statements. Question: Are value judgments tautologies? No—or at least *some* are not (and, be it noted, if 'sin is evil' is a tautology, it is not a 'good' one, not the kind that falls into the class of logical and mathematical propositions). Question: Are value judgments verifiable by observation? No: 'conceit is sinful' seemed to these observers to have no clear-cut linkage to an 'observable thing' definition base. Strictly, an ethical or aesthetic judgment is *meaningless*. Or it is at least 'cognitively' meaningless. What then are ethical or aesthetic utterances?

140

They are exclamations (or, if you prefer, ejaculations, impreca-
tions, or some such emotional 'expression'): they are expressions
of approving or disapproving feeling, affect, emotion. Such a
view was stated very sharply and simply by Ayer in the mid-
'thirties[4] and has been elaborated in varying degrees of detail
by others (rather fulsomely by Stevenson).[5] Let us assure ourselves as to what this view is really saying. As A. C. Graham,
an analytic philosopher, has recently pointed out, the view in
effect is saying that 'There can be no fruitful dispute over
questions of value except in terms of tastes and goals which
the disputants happen to share. Moses blew off his emotions
about murder by saying "Thou shalt not kill"; and if a killer
from Auschwitz happens to feel differently, *de gustibus non
disputandum*.'[6]

Within the ambience of analytic philosophy, the emotivist
view soon shaded over into (or was supplemented by) the
'imperativist' view, which holds that value judgments, pronouncements, standards, etc., are in effect commands. Since
analytic philosophers had stressed the multiplicity of 'use'-
functions in language, it was no longer to be maintained that
value utterances are *meaningless*, but rather that being commands, they 'behaved' in a very different way than did factual
statements. As Graham puts it: 'With this change of viewpoint
it becomes possible to admit that moral and aesthetic standards
do not say anything true or false about entities called "goodness" and "beauty" or about inclinations in taste, and yet to
hold they are as meaningful as statements of fact.'[7]

Analytic philosophy is currently extricating itself from the
almost incredible simplism of such positions in the work of
people like Hare,[8] Kerner,[9] and von Wright.[10] But it is too
optimistic to believe that views of this order no longer have cur-
rency in philosophy and, what is worse, they live on in large seg-
ments of the scientific community and of the culture at large,
into both of which contexts these views sifted, against little resistance, a long time ago. Indeed, there is a sense in which such
views existed in science and the general culture considerably
before they were reborn as technical philosophical formulations.
Of course, 'emotivism' and 'imperativism' must also be seen in
the line of subjectivistic views of value in philosophy and, to
some extent, psychology, of which they are perhaps the supreme

141

vulgarization. Subjectivistic theories agree in seeing value 'constituted' in relation to the needs, motives, purposes, interests, wants, wishes, of individual men or organisms: values are thus regarded as ends or goals of systems of motivation to which they are extrinsic in the very sense described in our earlier analyses of the 'extrinsic grammar.'

Though sustained analysis in the philosophic mode is not here possible, a few remarks may perhaps clarify certain consequences of the present position. Suppose we start with some such proposition as 'That picture is exquisite' or 'That action is noble.' It is hardly necessary to explain that from the present point of view these propositions are not equivalent (or fairly transformable) to 'Picture, wheee!' or 'Action, uhhuh!' (the 'emotivist' interpretation) or to 'Look at that picture!' (or 'buy it' or 'copy it'), or 'Emulate that action!' From the present point of view, such statements are just as much 'factual' as, say, 'That picture is large' or 'That action was rapid.' In some sense more 'factual' (if one be permitted degrees of 'facticity') in that explication, even at relatively gross levels, of the 'value properties' which '*in fact*' are governing discriminating application of 'exquisite', say in some concrete instance of a complex painting, could in principle convey far more information than explication of the characters by virtue of which the picture may be said to be 'large.' If the rejoinder be 'Yes, but there is still a difference in that virtually every user of a language could apply "large" correctly (and explicate the basis of its use), whereas in the case of "exquisite" we can expect much disagreement in application and in explication,' then I must ask the questioner to consider that the history of art and of taste provides overwhelming evidence that if we compose a hierarchy of language communities for 'exquisite' against appropriate criteria of training and sensitivity, we will find impressive intra-community agreements with respect both to application and explication.

One can only conclude that the concept of 'fact'—or its linguistic counterpart, as e.g., 'factual' or 'indicative' statement —is sorely in need of reanalysis. Facts must be in some way conceived as differentially charged with value, and whatever else a value may 'be,' it is a special kind of fact. Value statements in indicative form are not disguised exclamations or imperatives or, as other subjectivistic theories would have it,

142

SIGMUND KOCH

of interest, wish, or need. Indeed, it is rather more fruitful to regard value utterances in 'normative' or 'imperative' form as disguised *indicative statements*, in that, at least when made responsibly, they emerge from a context of knowledge or 'belief' within the speaker which could be unpacked into the 'factual' or 'indicative' statements on which the imperative is grounded.

The last point may be worth further examination. Let us concentrate on the case of a 'command.' When we assert a command, we do this within a context of belief that it will lead to such-and-such consequences for the commander, the commandee, for a third person, for a given group, the 'world,' etc. Part of that context of belief is a judgment that the intended consequent, or chain thereof, or ultimate consequent, is 'good,' 'bad,' 'desirable,' 'undesirable' (often in *specific ways*) with respect to such-and-such or so-and-so. Such a judgment can in principle be unpacked into a sequence of meaningful ('factually' or 'indicatively' meaningful, if one prefers) propositions *re* value properties. The mysterious 'leap' from the 'is' to the 'ought' is no mystery. It is no great epistemological profundity that the rules of English do not permit one to say that a 'command' is 'verifiable.' That it is useful for a language to contain specialized forms which register to a hearer that an action, change of belief, or some such thing is being called for, tells us no profound fact about the universe. As we have suggested, the context of knowledge, belief, expectation, etc., from which a command 'issues' can be described via a collectivity of 'is' statements, a sub-class of which can be said to be indicative of *value properties*. That 'is' often signalizes empirically 'verifiable' attribution (or predication, etc.), and 'ought' obligation (or an 'imperative' intent, or imperiousness, sententiousness, arrogance, petulance, prudishness, or passion) is a useful feature of language, but not a regulative principle of the universe.

To repeat, then, the concept of 'fact' needs radical revamping: 'facts' are suffused with 'value'; and 'value' diffused in 'fact.' The last statement is a homespun prolegomenon to a metaphysics. But as a *psychologist*, I can see no other beginning.

Turning now to the relations between values and science—as seen now by the *scientist*—it can be said that the simplisms long

143

accepted by many men of the white cloth exceed in vacuity even the philosophic ones which we have just examined. Thus, it is still something like standard belief that science is concerned exclusively with 'neutral' factual relationships, that it seeks 'objective' knowledge of invariable associations of observable events, or probable empirical regularities, that it can talk about instrumental relationships or 'means-end' relationships, *but can say nothing about ends*. In so far as the standard patter permits reference to 'values,' their relevance must be allocated, so to say, to the human or social context in which, for some obscene reason, science happens to be embedded. Thus it is permissible (at least on the part of the tender-minded) to suggest that the scientist, as a *citizen*, should be concerned with the relation between scientific knowledge and human welfare, that it is desirable that he take an interest in determining that scientific knowledge be applied to 'good' social ends and not evil ones, etc. The *very* daring are even now beginning to go so far as to admit that certain moral traits of the *scientist* have a relevance to science—that it is desirable that he have a respect for the 'truth,' desirable that he not fudge his data.

Perhaps the reader will feel with some indignation that the preceding account is too cynical: is there not, after all, a great clamor now shaping up in scientific circles about such matters as the 'responsibility of the scientist'? There in fact is, and I would be the last to think it an unhappy development. But *even* those currently concerned to bring about sharper recognition of the relevance of value factors to science tend to proceed as if the issues can be encapsulated and dispatched in two or three generalizations, or perhaps admonitions. A recent case in point may be found in C. P. Snow's article on 'The Moral Un-neutrality of Science' in which the story comes down pretty much to the fact that scientists respect the truth, and should; that they tend more than most groups to care about people, inclusive of the poor, and should; that they lead relatively cleaner lives with less divorce than do, on the average, members of certain other groups, and should.[11] From another of Snow's contributions, we learn that scientists have 'the future in their bones.'[12] And should.

In rather marked contradistinction to the stereotypes concerning science-value relations, the present position suggests considerations of the following order: Value-property distributions will suffuse all perception, all meaning; such 'parameters' of psychological process do not hygienically cancel out when the context becomes one of science. Scientific languages, no matter how restricted, will generate meanings suffused by value properties; indeed, the languages themselves considered as perceptual objects will also 'generate' value properties.

Much of science is concerned with the understanding, explication, or causal analysis of value phenomena—or perhaps more properly, phenomena suffused to one or another extent with value properties. If one were to make an extremely global generalization (requiring marked qualification in certain areas), it could be said that as one ascends the scale from the physical sciences through the biological sciences to the social sciences and finally the humanities, one confronts 'subject-matters' more and more richly permeated with value properties or, more precisely put, these various subject-matters may be ordered in terms of differential concentration upon value-property aspects of each. If it be maintained that those who work at the *scientific* levels are *confined* to *factual* statements about the value 'aspects' of their subject-matter, then I reply that we have shown conventional usage of 'fact' to be either incorrect or systematically ambiguous in the very sense that begs the question. Surely if a scientist discovers a polio vaccine, no semantic prohibition should prevent him from 'leaping' from a well-formed and well-evidenced means-end statement to the 'ought'-injunction that adds a directive-rider to the cognitive content. If it be maintained that the scientist is still merely making a conditional statement of the form 'if you wish (for some damn fool reason) to remain well . . ., etc., etc.,' then the uncouth answer might well be that the state of health is self-recommending by virtue of intrinsic value properties.

Consider now the *actual human context* of science. Or better, the human center. Put the scientist, the actor, back in the picture from which he has been for so long excluded. We have argued that value-determining properties are generated by the neural processes which are the substrate of action; that they must in some sense be thought criterial with respect to the moment-by-moment directionality of action. That we can as yet only speak a metaphorical and vague language about such matters has nothing to do with the force of the evidence that recommends such metaphors. In reaching toward understanding, lawful analysis, 'control,' if you will, of the phenomena which entice their interest, scientists are reaching toward meanings, toward the maximization of certain value properties immanent within those meanings, and throughout are guided by value-determining properties which are some function of the concurrent 'input' and their own internal 'processings.' The latter will reflect, among other factors, those quasi-permanent 'structures' that we also sometimes call 'values' in the sense manifested by durable preference-dispositions, etc. Values in this sense are often roughly equated with 'tastes,' 'prizings,' 'attitudes,' sometimes with 'needs,' etc. However such enduring value systems be conceptualized, they may be seen as organizations which govern the moment-to-moment salience, potency, of differential 'clusters' or ranges of 'specialized' value properties.

Such durable value dispositions (call them Values) will be heavily implicated in the 'options' that a scientist asserts in inquiry, the moment-to-moment decisions that he makes. Are such events rare? There is a way of describing inquiry which sees it, and validly so, as a sequence of human options, a flow of decisions. Even the obvious Value foci provided by the optional elements in science tend to be by-passed in standard discussions of science and value. Consider the contingency, at all times, of the scientific enterprise on the 'aims' of the scientist, on his predilections with respect to choice of 'method,' both conceptual and empirical, his predilections with respect to mode of problem formulation, his sense of scientific 'importance,' his perception of the relation between his actions and the standards of his colleagues or the needs of society; his predilections concerning factors in light of which he adjusts his scientific beliefs or assertions to 'evidence'; his preferences with respect to modes of theoretical processing, with respect to optimal modes of conduct in the polemical situations of science, and on and on. Hopefully, many of his options are fixed upon in terms of the 'rational' productivity of their consequences in the choices of other men, present and past. Hopefully, they remain contingent in some degree upon such consequences as experienced in his own biography. But such matters are *choices*; they are *decisions*. Nothing should conceal their determination by processes having

146

marked value components (however these be phrased), nor should anything conceal the value aspects of the web of commitment and dedication which interpenetrates all the activities of scientists.

Paul Dirac has recently been widely quoted as having written, 'It is more important to have beauty in one's equations than to have them fit experiment.' Polanyi has placed strong and consistent emphasis on what he has called the 'harmonious character' of a theoretical conception or set of ideas in guiding its creation, elaboration, productiveness, and fate. Virtually all major creative figures in the history of theoretical physics and of mathematics—and before these, the Pythagoreans—have said something of the same sort. Indeed, it is a *truism* of the 'formal' disciplines generally—logic as well as mathematics—that 'elegance' is a prime desideratum. And even logical positivists have included on occasion 'elegance'—along with 'economy' and the like—as a mark of the 'fruitfulness' of theory. Can this vast range of agreement point to something purely adventitious? At this inordinately complex and important level too, then, value is criterial with respect to science.

The present analysis may have seemed at times to be absorbing *everything* in science, all of 'fact,' into value. Perhaps it has in some sense gone too far in this direction: it is not yet easy to talk about such problems. But judging from the circumstances just mentioned, the present account is in good company. What I should like to urge, though, is that, instructive though it be to acknowledge 'beauty,' 'harmony,' 'elegance,' as criterial with respect to scientific theory, to the processes of thinking which result in theory, this is but the beginning of instruction. 'Beauty,' 'harmony,' 'elegance' land us back among the most global value-abstractions of aesthetics; they are on a par as well with 'good' and 'right' in ethics. The burden of this paper has been to urge that the human race cannot for ever rest content with so non-specific and gross a level of analysis of its most meaningful accomplishments, prized artifacts, significant phenomena. The concept of value properties may involve no more than an assertion of faith that finer degrees of specification are possible. Yet in our most fluent dialogues with ourselves and with our friends, when they are fluent, we have proved on countless occasions that finer knowledge is possible.

147

REFERENCES

[1] I have dealt more laboriously with the historic phases of behaviorism in 'Clark L. Hull,' in *Modern Learning Theory* (W. K. Estes *et al.*), New York: Appleton-Century-Crofts, 1954, pp. 1–176; 'Epilogue,' in *Psychology: A Study of a Science* (S. Koch, ed.), New York: McGraw-Hill, 1959, pp. 729–88; 'Behaviorism,' in *Encyclopaedia Britannica, 3* (1961), pp. 326–9; 'Psychology and Emerging Conceptions of Knowledge as Unitary,' in *Behaviorism and Phenomenology* (T. W. Wann, ed.), Chicago: University of Chicago Press, 1964, pp. 1–41.

[2] G. A. Miller, E. Galanter, and K. H. Pribram, *Plans and the Structure of Behavior*, New York: Henry Holt, 1960.

[3] S. Koch, 'Behavior as "Intrinsically" Regulated: Work Notes Toward a Pre-theory of Phenomena called "Motivational",' in *Current Theory and Research in Motivation, 4* (M. R. Jones, ed.), Lincoln: University of Nebraska Press, 1956, pp. 42–86.

[4] A. J. Ayer, *Language, Truth and Logic*, Oxford: Oxford University Press, 1936.

[5] C. L. Stevenson, *Ethics and Language*, New Haven: Yale University Press, 1944.

[6] A. C. Graham, *The Problem of Value*, London: Hutchinson University Library, 1961, pp. 13–14.

[7] *Ibid.*, pp. 14–15.

[8] R. M. Hare, *Freedom and Reason*, New York and Oxford: Oxford University Press, 1963.

[9] G. C. Kerner, *The Revolution in Ethical Theory*, New York and Oxford: Oxford University Press, 1966.

[10] G. H. von Wright, *The Varieties af Goodness*, London: Routledge & Kegan Paul; New York: The Humanities Press, 1963.

[11] C. P. Snow, 'The Moral Un-neutrality of Science,' *Science, 133* (1961), no. 3448, pp. 256–9.

[12] C. P. Snow, *The Two Cultures and the Scientific Revolution*, New York: Cambridge University Press, 1959.

6

THE MESSAGE OF THE HUNGARIAN REVOLUTION

Michael Polanyi

In February 1956, Krushchev denounced the insane regime of Stalin at the 20th Party Congress. Four months later, in June 1956, the repercussions of this act led to an open rebellion of the Hungarian writers at the meetings of the Petöfi Circle in Budapest. This was the actual beginning of the Hungarian Revolution, which broke out violently in October of the same year.

It has often been mentioned, but never realized in its full importance, that the rebellion of the Petöfi Circle was an uprising of Communist Party members. There were many among them who had won high honors and great financial benefits from the government. They had been, until shortly before, its genuine and passionate supporters.

The Petöfi Circle, where this first rebellion broke out, was an official organ of the Party, presided over by trusted representatives of the ruling clique. But in June 1956 other Party members, forming the bulk of the audience, overcame the platform. They demanded a reversal of the position assigned to human thought in the Marxist–Leninist scheme. Marxism–Leninism taught that public consciousness is a superstructure of the underlying relations of production; public thought under socialism, therefore, must be an instrument of the Party controlling Socialist production.

The meetings rejected this doctrine. They affirmed that truth must be recognized as an independent power in public life. The press must be set free to tell the truth. The murderous trials based on faked charges were to be publicly condemned and their perpetrators punished; the rule of law must be restored. And, above all, the arts corrupted by subservience to the Party must be set free to rouse the imagination and to tell the truth.

It was this outbreak that created the center of opposition which later overthrew the Communist government of Hungary.

Has the response of the West to these events been adequate? I am not asking whether we ought to have aided the Hungarian Revolution by money or the force of arms. I am asking about our intellectual and moral responses.

What did we think, what do we think today, of this change of mind among many of the most devoted Hungarian Stalinists? Do we realize that this was a wholesale return to the ideals of nineteenth-century liberalism? In the Hungarian Revolution of 1956 its fighters were clearly going back to the ideals of 1848, to Liberty, Equality, Fraternity. After the French Revolution the ideas of liberty had filtered through the whole of Europe, and in 1848 these ideas aroused a chain of insurgences. Starting from Paris, rebellions swept through the German states and through Austria into Hungary. The Hungarian revolutionaries of 1956 revived these same ideas; they believed that the ideals of truth, justice, and liberty were valid and they were resolved to fight and conquer in their name.

I ask you again: What do we say to this? It has been a long time since history was last written in the English language as the inevitable progress of truth, justice, and liberty. I think J. B. Bury was the last to do so fifty years ago. Our current interpretations of social change and of the rise of new ideas are nearer to the theories of Marx than to the views of Bury, or those of J. S. Mill, or of Jefferson, or Condorcet and Gibbon. Our interpretations are couched in terms of a value-free sociology, of an historicist historiography, or of depth psychology. We can then hardly be expected to acknowledge the aims which the Hungarian rebels professed to be fighting for. Our most popular explanation of unrest in the Soviet countries is sociological. It says that the methods of Stalin had fulfilled their purpose of carrying out the rapid industrialization of the Soviet Union, but that these methods were no longer suited to a more complex society and hence the need had arisen for new ideals, more appropriate to the new situation. This is said to have caused the renewal of thought in the Soviet countries since Stalin's death.

Such theories are put forward without any supporting

evidence, and nobody asks for evidence. It is ignored that two of the most highly industrialized regions of Europe, namely Eastern Germany and Czechoslovakia, have produced the hardest Stalinist governments and resisted reform the longest. Any mechanical explanation of human affairs, however absurd, is accepted today unquestioningly.

Listen to the story told three years ago in *Encounter* by a distinguished expert on Soviet countries, Professor Richard Pipes, then Associate Director of Harvard's Russian Research Center. Professor Pipes wrote:

Four years ago, when writing an essay on the Russian intelligentsia for the journal *Daedalus*, I wanted to conclude it with a brief statement to the effect that the modern Russian intellectual had a very special mission to fulfill: 'to fight for truth.' On the advice of friends I omitted this passage since it sounded naive and unscientific. Now I regret having done so. . . .[1]

Thus at the most influential academic center studying Soviet affairs, it took three years after the rebellion of writers at the Petöfi Circle for it to be mooted for the first time that this kind of unrest was due to craving for truth. Even then this suggestion was suppressed in deference to expert opinion, because to speak of intellectuals fighting for truth was held to be 'naive and unscientific.'

Professor Pipes continues this passage by explaining why he thinks now (four years later) that in the case of Russian intellectuals it is not unscientific to attribute to them a craving for truth. The reasons for this apply equally to the conditions under which Hungarian intellectuals rebelled, and thus the argument applies to them too. Professor Pipes writes:

The reason the word 'truth' is in disrepute among us is because we attach to it generally moral connotations; that is, we understand it as a concept which implies the existence of a single criterion of right and wrong—something we are not willing to concede. We react thus because in the environment in which we live our right to perceive is not usually questioned; what can be questioned is our interpretation of the perceived reality. But in an environment where the very right to perception of reality is inhibited by claims of the State, the word 'truth' acquires a very different meaning. It signifies not true value but true experience: the right to surrender to one's impressions without being compelled for some extraneous reason to interpret and distort them.[2]

We are told here that it is unscientific to speak of 'truth' in our society and that the word merits its current discredit because we attach moral values to it. It is legitimate to recognize that the Soviet intellectuals fight for truth because they are demanding merely the right to perceive reality and not to convey any interpretation of it in terms of values believed to be true. However, the writers who rebelled in the Petöfi Circle ten years ago were not demanding the right to have certain 'perceptions of reality.' They demanded that the execution of Rajk, based on faked charges, be publicly admitted and that they be free to denounce the tyranny forcing its false values on the people; they were demanding the right to write the truth in novels, in poetry, and in the newspapers.

I think that to understand this rebellion we must see it as part of a major historical progression, composed of three stages to be seen jointly. The first was the defection of leading intellectuals of the West, abandoning 'the God that failed.' Some turned away during the trials of 1937–38, others after the Stalin-Hitler pact of 1939, others still later. By the end of the Second World War there was an ardent band of former Communists desperately warning against the Soviets, whose philosophy made them impervious to our values.

The next decisive change occurred in Soviet Russia itself on the day Stalin died. The first act of his successors was to release the thirteen doctors of the Kremlin who had recently been sentenced to death on their own false confessions of murderous attempts against Stalin and other members of the government. This action had a shattering effect on the Party. If Party-truth was to be refuted by mere bourgeois objectivism, then Stalin's whole fictitious universe would presently dissolve.

The alarm was justified. After the further revelations of the 20th Party Congress, it became clear that the new masters of the Kremlin had acted as they did because they felt that their position would be safer if they had more truth on their side and less against them. They sacrificed the most powerful weapon of terror for this end—the weapon of faked trials; no such trial was held in the Soviet empire after the Jewish doctors were set free. A radical change of the foundations had begun.

The third major step in the growth of the power to vindicate the truth—and the values resting on the right to tell the truth—was the rebellion of Hungarian writers. We can see this now as part of the historical process which had started with the numerous defections from Communism among Western intellectuals about fifteen years earlier. The rebellion of the Petöfi Circle was a change of mind by dedicated Communists forming an intellectual *avant-garde*. Though undertaken against the still ruling Stalinism, it spread under the very eyes of the secret police, until one day it brought the workers out into the streets and won over the army to its support. The government was overthrown. And, though a few days later the Russians moved in with new tanks, they did not restore the Stalinists to power. Since October 23, 1956, ideas have never ceased to circulate freely in Hungary. You cannot fully print them, but you can spread them effectively by word of mouth.

When seen in this context these events might appear self-explanatory. They would indeed be so if the truth of the Enlightenment appeared self-evident today. But though these truths still dominate our public statements and largely dominate also the day-to-day judgments of men, they are, for the most part, ignored by social scientists. Most academic experts will refuse to recognize today that the mere thirst for truth and justice has caused the revolts now transforming the Soviet countries. They are not Marxists, but their views are akin to Marxism in claiming that the scientific explanation of history must be based on more tangible forces than the fact that people change their minds.

Such a split between the laity and the learned is not unusual in the history of thought, and has at times been beneficent. But I think that the major responsibilities of our time require a clarification of these inconsistencies. I have quoted the hesitations of Professor Pipes of Harvard and have criticized the advisers mentioned by him for their reluctance to acknowledge the fight for truth among Soviet intellectuals. I will now take a look at the principles governing this attitude. I shall use as my source a distinguished textbook entitled *Politics and Social Life: An Introduction to Political Behavior*, published in 1963.[3]

This textbook, composed of seventy contributions by well-known academic scholars, covers 830 large pages. It claims to teach the 'Science of Social Behavior,' both its principles and its application to several subjects. The introduction, by Robert A. Dahl of Yale University, declares that the proper subjects of science are 'events which may be observed with the senses and their extension.' This excludes from science any statements of value. Values admittedly enter into the *lives* of social scientists, but in the pursuit of science the scientist makes only statements of fact that can be observed by our senses. To state, for example, that the Supreme Court of the United States has nine members is to state a fact, while to say that its decisions are impartial is not to state a fact; this kind of statement is to be excluded from the science of political behavior.

Professor Dahl follows up these principles by an attack on traditional political science. He says—and in this he is right—that this kind of inquiry is concerned with such questions as 'What is a good society?' He derides such a pursuit by calling it mere 'political theology.'

To see how these scientific principles work in our case, I shall now quote some detailed evidence of the mental transformation among Hungarian Communists and other Soviet intellectuals. Here is a passage written at the beginning of October 1956 (that is, a few weeks before the outbreak of the revolution), by Nicholas Gimes, until lately a firm Stalinist:

> Slowly we had come to believe, at least with the greater, the dominant part of our consciousness . . . that there are two kinds of truth, that the truth of the Party and the people can be different and can be more important than the objective truth and that truth and political expediency are in fact identical. This is a terrible thought . . . if the criterion of truth is political expediency, then even a lie can be 'true' . . . even a trumped-up political trial can be 'true'. . . . And so we arrived at the outlook which infected not only those who thought up the faked political trials but often affected even the victims; the outlook which poisoned our whole public life, penetrated the remotest corners of our thinking, obscured our vision, paralysed our critical faculties and finally rendered many of us incapable of simply sensing or apprehending truth. This is how it was, it is no use denying it.[4]

Speaking of faked trials, Gimes must have had foremost in mind the trial and hanging of Laszlo Rajk. The grave of Rajk

154

was visited at about this time by a procession of many thousands, led by Communists. The pilgrimage took place on October 6, the day that traditionally commemorates in Hungary the hanging (at the orders of the Austrian Emperor) of thirteen revolutionary generals in 1849.

Let me add here some extracts from a later (non-Hungarian) document. It is a long poem by a leading Bulgarian Communist, Dimitar Metodiev, published in the official Party organ of Bulgaria in January 1962.[5] It describes in passionate terms the great change that by this time was universally imposed by the Party. The poem, entitled 'Song of Our Faith,' starts with the verse:

> If one had told me yesterday:
> —Friend take off your glasses!—
> I would have laughed
> in a black rage
> at the mortal insult.

and it continues later:

> On the ground now
> The glasses lie smashed
> And I look at them stunned.
>
> Friends!
> the great hour of the twentieth congress
> has struck our earth.
> Oh let its holy justice
> solemnly, inexorably go forth today.

Returning once more to Hungary, look at a case in which 'the smashing of the glasses' happened, long before its official enforcement by the Party and even before any public manifestations by the Hungarian intellectuals. It happened to the Hungarian Communist writer Paloczi-Horvath, while imprisoned at the orders of Rakosi, as described in his book *The Undefeated*.[6] Since his arrest in September 1949, Paloczi had suffered incessant cruelties, inflicted on him to make him confess to false charges; yet he had never wavered in his Communist faith. Then suddenly, after two years in prison, in a matter of a few days, he changed his mind and rejected the Party. His sufferings in prison continued unchanged, but from this moment he felt free, even happy—a new person.

Let me now confront the science of political behavior with these pieces of evidence and see what such a science can make of them. Let us take a generous view of the range included among 'events which may be observed by the senses and their extension.' Thus we may accept it as a scientifically observed fact that two very different ways of seeing political affairs have been current in the Soviet countries: one until Stalin's death in March 1963 and another from about two years later, after the 20th Party Congress. A pair of hitherto indubitable spectacles were smashed, and everything looked different from then on. Science can record this fact.

But does it account for *the act of smashing the spectacles*? Does it explain this act, which in Hungary amounted to a revolution? Our science could acknowledge this fundamental change of outlook as a fact—and perhaps even class it as an intellectual achievement—if the change could be regarded as the correction of a hitherto current theory by discovery of new factual evidence. But this is not what happened. Paloczi-Horvath acknowledges with amazement that all the facts he was seeing in a new light after his defection from Communism had already been known to him as facts before his change of mind. Arthur Koestler has described this phenomenon in looking back on his time as a Communist. 'My party education had equipped my mind [he wrote] with such elaborate shock-absorbing buffers and elastic defences that everything seen and heard became automatically transformed to fit a preconceived pattern.'[7] Stalinism was a closed system, a system of the kind I have described in *Personal Knowledge* as capable of accommodating any conceivable new piece of evidence. Such a system (one that can interpret in its own terms any possible fact) can be shaken only by preference for a total change of outlook. The Bulgarian Dimitar Metodiev was right in describing it as the smashing of one's spectacles.

Remember that the whole person is involved in trusting a particular comprehensive outlook. Any criticism of it is angrily rejected. To question it is felt to be an attack on an existential assurance protected by the dread of mental disintegration. We have many testimonies by Communists, that their minds clung to Marxism–Leninism for fear of being cast into the darkness of a meaningless, purposeless universe. Paloczi-Horvath himself

156

describes how, after his abandonment of the Communist faith, he at first felt lost, incapable of coherent thought. It was only after a few days that he began to experience a new way of making sense, which he felt to be a more honest, decenter way of thinking, bringing him a happiness that transcended his sufferings in prison.

Nicholas Gimes confirms this emotional value of a comprehensive framework. Looking back on his time as a devoted member of the Party when he believed that advantage to the revolution was the criterion of truth and of all other human values, he feels horrified by the mental corruption caused by this doctrine. The scornful rejection of objectivity as a bourgeois delusion and fraud, which he had previously accepted as an intellectual triumph, now appears to him as an abysmal loss, the loss of the capacity to see any truth at all.

We have many descriptions also of conversions in the opposite direction. Thousands of leading intellectuals have joined the Communist Party because they found in Marxism an interpretation of history and a guide to action far superior to the outlook predominant since the French Revolution among progressive circles in the West. In 1917, when President Wilson was confronted by Lenin on the world scene, these people laughed Wilson's language to scorn and dedicated themselves instead to Lenin's doctrine, which they felt to be intellectually and morally superior.

We may take it, then, as an established fact that during the past half-century radical changes of mind have taken place, which were motivated each time by passionate affirmations of value. And we can return again, with these facts before us, to the science of political behavior and ask how this science deals with such conversions from one basic conviction to another.

I shall admit once more the widest possible interpretation of this scientific method, and shall concentrate on the question of how the method treats the moral motives of men when they smash their spectacles and put on new spectacles instead.

The behavioral scientist acknowledges that moral values motivate him in private life. He may meaningfully observe that moral motives are claimed by people for their actions. Such reference to alleged values was sanctioned by Rickert in 1902 in his book *Die Grenzen der Naturwissenschaftlichen Begriffs-*

bildung.[8] When he taught here that value judgments are beyond the competence of science, he excluded *Werturteile* from science but allowed that *Wertbezogenheit* is observable by science. Through the mediation of Max Weber, this doctrine has since been used to reconcile moral neutralism in science with observing the experience of moral motivation in persons studied by the scientist. According to this theory, the scientific method can establish, for example, that the rebellious Hungarian intellectuals attributed moral value to the freedom of truth and felt that the doctrine of Party-truth caused a moral corruption of public life; but it cannot observe whether this moral judgment was right or not, since this observation would constitute a moral judgment which science is not competent to make.

However, this doctrine can be proved to be inconsistent by two complementary arguments, one starting from the existence of the scientist as a moral agent (outside the range, as he alleges, of his scientific activity) and examining the implications of his (extra-scientific) moral judgments for the statements he makes as a scientist; the other moving from his allegedly value-free scientific pronouncements to his (extra-scientific) moral judgments.

The first argument is as follows. (1) All men, whatever their professions, make moral judgments. (2) When we claim that an action of ours is prompted by moral motives, or else when we make moral judgments of others—as in recognizing the impartiality of a court of law—we invariably refer to moral standards *which we hold to be valid.* Our submission to a standard has universal intent. We do not prefer courts of law to be unbiased in the same sense in which we prefer a steak to be rare rather than well done; our appeal to moral standards necessarily claims to be *right*, that is, binding on all men.[9] (3) Such a claim entails a distinction between *moral truth* and *moral illusion.* (4) This distinction in turn entails a distinction between two types of motivation. The awareness of moral truth is founded on the recognition of a valid claim, which can be reasonably argued for and supported by evidence; moral illusion, in contrast, is compulsive, like a sensory illusion. (5) Thus once we admit, as we do when we acknowledge the existence anywhere of valid moral judgments, that true human values exist and that people can be motivated by their know-

158

MICHAEL POLANYI

ledge of them, *we have implicitly denied the claim that all human actions can be explained without any reference to the exercise of moral judgment.* For to observe—in our present case—that the Hungarian writers rebelled against the practice of faked trials because they *believed* this practice to be wrong, leaves open the question *why* they believed this. If true human values exist, the Hungarians may not have been driven by economic necessity, propaganda, or any other compulsion; they may have been rebelling against a *real evil*, and may have done so *because they knew it to be evil.* But this cannot be decided without first establishing whether faked trials are in fact evil or not. Therefore: (6) *This value judgment proves indispensable* to the political scientist's explanation of their behavior.

The inconsistency of a science professing that it can explain all human action without making value judgments, while the scientist's private actions are said to be often motivated by moral motives, can be more simply demonstrated the other way round. If the social scientist can explain all human actions by value-free observations, then none of his own actions can claim to be motivated by moral values. Either he exempts himself from his own theory of human motivation, or he must conclude that all reference to moral values—or any other values—are meaningless: are empty sounds.

Such a conclusion, though repulsive, would at least appear consistent.[10] But there exists another consistent position that makes better sense, namely the kind of position affirmed by the Hungarian revolutionaries. They expressed their horror of a philosophy which taught that truth, justice, and morality must be subservient to the Party, and demanded that these values be recognized as free powers of the mind. Their moral motives were confirmed by their claim that the standards of truth, justice, and morality must be recognized as independent powers in public affairs. A passion for freedom thus confirms the existence of the freedom it fights for.

This analysis shows that a science that claims to explain all human action without making a value judgment not merely discredits the moral motives of those fighting for freedom, but also discredits their aims. This is why the Hungarian revolutionary movement, which revived the ideals of 1848 and which claimed that truth and justice should be granted power over

159

public affairs, has met with such a cold reception by the science of political behavior. Modern academic theories of politics give support, on the contrary, to the doctrine which denies that human ideas can be an independent power in public affairs.

It goes without saying that these implications of their science were contrary to the sympathies of the scholars wedded to this science. It is true that Professor Pipes hesitated for three years before he first tried to state the obvious fact that Soviet intellectuals were fighting for truth. But when he was stopped from doing so by his more fastidious colleagues, he did not give up; after another four years he produced a labyrinthine theory which allowed him to state this fact.

Such scholarly prevarications may appear harmless, but they are not, for they obscure great events which dominate our history. No vision of these events can then arise to guide us. We are bogged down in meaningless details. Think of the great figures of the Enlightenment. Think of Condorcet hiding in Paris from the Jacobins and completing before his death his triumphant forecast of universal progress. And then look at the article in the volume on the science of political behavior which deals with 'Soviet Domestic Politics Since the Twentieth Party Congress,' an article originally published in 1958.[11] Its author expects to see a growing influence of indoctrination in Russia on the grounds that extreme measures to secure ideological conformity, like those of the Zsdanov period, have been given renewed prominence since mid-1957. This is the kind of hand-to-mouth foresight which the science of social behavior has produced, for lack of vision of what is actually going on.

Remember now the contemptuous rejection by the science of social behavior of such questions as 'What is a good society?' Such inquiries, belonging to traditional political science, were derided as 'social theology.' But what the revolution in Hungary—and the whole movement of thought of which it formed a part—have declared, was a doctrine of political science in the traditional sense of the term. The movement condemned a society in which thought—the thought of science, morality, art, justice, religion—is not recognized as an autonomous power. It rejected life in such a society as corrupt, suffocating, and stupid. This was clearly to raise, and to answer up to a point,

160

the question 'What is a good society?' It was to establish a doctrine of political science.

Actually the current transformation of thought in the Soviet countries had added to political science new and remarkable insights. We have seen that this social transformation was achieved by a smashing of spectacles. Its typical utterances, such as those I have quoted, manifest the deep emotional upheaval caused by recognizing once more that truth, justice, and morality have an intrinsic reality; and this is the decisive fact. Paloczi-Horvath remained in prison, but his conversion made him free; and that symbolizes the character of this whole immense political movement. It was not started by demands for institutional reform, and it went on without materially changing the institutional framework of the countries it transformed.

I know how very peculiar was this transformation of the Soviet countries, for I have often been challenged about it during the past years. When I spoke of the radical changes going on in the Soviet countries, I was faced sharply with the objection that these movements had brought no constitutional changes and could therefore be reversed at any moment. But this argument misconceives the foundations of power in this age ruled by rival philosophies. I myself yielded in a way to this error when, at the beginning of this talk, I called Stalin's regime 'insane.' His regime was actually well founded on the strange assumption that neither truth nor justice, nor art, nor human feeling exist except in the service of the Party and subject to its rule. This conviction went so deep, was so fully accepted even by its victims, that when faithful Communists were executed on trumped-up charges, they went to their death shouting 'Long live Stalin!' And this logic works also in reverse: Stalin's regime virtually ceased to exist when its basic conceptions of intellectual and moral reality lost their hold on thought.

But we may yet ask, why trust this change of basic conceptions? Could people not go back to their previous outlook, which supported Stalin for so long? There is indeed nothing to prevent them from doing so, except the fact that their present conception of mental reality is far nearer to the truth, and they feel this and enjoy it. In this sense, and in this sense alone, can we say that what has happened has not been merely a *changing* of spectacles, but a *smashing* of spectacles. And this is a change

161

more radical and more lasting than any constitutional reforms could be.

We have seen that the vagaries of philosophy, which have hardly ever been followed literally among men before, often control our fate today. The Soviet people have gone through a philosophical experiment which revealed to them the nature of mental reality in a way not recognized in our universities. They are turning away from the site of their calamities and seek support in the West. This we fail to give them. Our scientific methods search for truth by turning a blind eye on the values which the rebellious Communists have recognized, and our traditional schools of political science fail to take the chance that history has given them to vindicate these values. They are disabled by the anti-metaphysical pressure surrounding them.

Meanwhile, in the Soviet countries, the ideas of mental independence are not only renewing the grounds of thought, but are daily revealing new sights previously blocked by official fictions. Hungarian newspapers have published complaints by Party members that scant respect is paid to them. They say that the government itself prefers non-Party members. We can understand what is happening here. Those who still hotly profess the Marxist interpretation of history, who proclaim the surpassing achievements of revolutionary socialism and predict the proximate coming of Communism—these people are increasingly looked upon as ignorant fools. The young are immune against their teachings. The Old Believers' authority is drained away by the superior intelligence surrounding them. Everywhere outside China this change is clearly in progress.

The change is bound to be accelerated by the gradual acceptance, since 1963, of Liberman's critique of detailed planning as inferior to market operations controlled by profit. It is becoming clear that there is in fact no viable socialist alternative to the market. Thus the Russian Revolution, which had conquered power in order to achieve a radically distinct form of economic organization, one that would be far more productive and also morally superior to commercial management, has now demonstrated the fact that there is no such possibility. By the time its fiftieth anniversary is celebrated, the Revolution may be widely recognized by its very successors as having been

162

virtually pointless. It may live on henceforth in its emblems, as the Mexican Revolution has lived on, without making much difference in substance.

On the fortieth anniversary of the Russian Revolution, close on nine years ago, I published an article in *Encounter* under the title 'The Foolishness of History.' I complained that our scientific approach to history debars us from envisaging the part of folly in shaping history. I wrote that:

. . . volume upon volume of excellent scholarship is rapidly accumulating on the history of the Russian Revolution, but as I read these books I find my own recollection of the event dissolving bit by bit. . . . The Revolution is about to be quietly enshrined under a pyramid of monographs.

Yet I know that it was something quite different. Not only when it actually happened; but all along, up to this day—for it still lives in our own blood. It was boundless; it was infinitely potent; it was an act of madness. A great number of men—led by one man possessing genius—set themselves limitless aims that had no bearing at all on reality. They detested everything in existence and were convinced therefore that the total destruction of existing society and the establishment of their own absolute power on its ruins would bring total happiness to humanity. That was—unbelievable as it may seem— *literally* the whole substance of their projects for a new economic, political, and social system of mankind.[1][2]

I think that this is now dawning on the minds of people all over the planet. The realization of the fact that we have wasted half a century of European history and in the process well-nigh destroyed our civilization, may lead to terrible thoughts beyond my horizon here. On the other hand, the final dissolution of the bogus salvation promised by our revolutions may exorcise at last the bogy of its hellish powers. Once the disasters of the past fifty years are clearly seen to have been pointless, Europeans may turn once more to cultivate their own garden. Such a new flowering of Europe may yet save this planet from the familiar mortal danger threatening us all today and for all future times to come.

REFERENCES

[1] R. Pipes, 'Russia's Intellectuals,' *Encounter*, 22 (Jan. 1964), pp. 79–84, p. 83.
[2] *Ibid.*

[3] N. W. Polsby, R. A. Dentler, P. S. Smith (eds.), *Politics and Social Life*, Boston: Houghton Mifflin, 1963.

[4] N. Gimes, in a Budapest periodical, October 3, 1956.

[5] D. Methodien, 'Song of Our Faith,' *East Europe*, *11* (April 1962), pp. 8–11.

[6] G. Paloczi-Horvath, *The Undefeated*, Boston: Little, Brown, 1959.

[7] A. Koestler, *The God That Failed*, London: Hamilton, 1950, p. 68.

[8] H. Rickert, *Die Grenzen der Naturwissenschaftlichen Begriffsbildung*, Tubingen: Mohr, 1902.

[9] This claim is not impaired by the fact that the application of true values may be subject to exceptions. Arsenic is rightly known as a poison, and aspirin as a harmless drug, though some people are immune to arsenic and every year a few people die from taking a tablet of aspirin.

And let me note that, contrary to Kant, the universality of moral standards applies only to the judgments of men and not to their behavior for men's situations are often incommensurable.

[10] Once it is acknowledged, however, that *any* claim to truth entails a recognition of standards, even the scientist's 'value-free' statements of fact are seen to be unable to claim their own truth in light of his own theory. For the purpose of the present argument, however, it is not necessary to follow through the epistemological implications of the doctrine to this ultimate self-contradiction.

[11] See note 3, above.

[12] M. Polanyi, 'The Foolishness of History,' *Encounter*, *9* (November 1957), pp. 33–7, p. 3.

7

BEING AND DOING:
A STUDY OF STATUS RESPONSIBILITY
AND VOLUNTARY RESPONSIBILITY

John R. Silber

The doctrine of original sin, according to which man exists in a fallen state, in a diseased condition, having begun his earthly career under a burden of guilt derived from Adam, is repugnant to most contemporary minds. Its repugnance derives not from its mythological elements which, by a Bultmannian re-section, might be cut away, but from its foundation in and dependence on a notion of *status*[1] *responsibility*: it finds man morally wrong and morally blameworthy for his diseased *condition* or *state of being* rather than for any specific *conduct* of a morally blameworthy character. The doctrine of original sin is radically at odds, therefore, with the legal definition of criminal conduct and the contemporary philosophical view of moral obligation, both of which are based on a notion of *voluntary responsibility*. According to the latter notion, man is morally wrong and blameworthy, not for what he *is* but for what he *does*. The conceptions of status and voluntary responsibility are both ancient, both enshrined in mythological lore. Yet it is a curious and important historical fact that the conception of voluntary responsibility has become dominant in both ethics and in criminal law, while the conception of status responsibility has scarcely survived. The latter conception is generally rejected by ethicists as being itself an immoral notion; and while it is not totally rejected by criminal courts, lawyers, and legal theorists, it is accepted into criminal law only under severe limitations, with grave doubts over vigorous opposition, and for extra-moral considerations.

Following a very brief survey of law and ethics to introduce some empirical support for these observations, I shall try to show that while neither the voluntary nor the status conception of responsibility is satisfactory by itself, both are required with

M   165

modifications in the formulation of a sound theory of responsibility. I shall argue that recent tort law provides useful guidelines for the extension of the concept of *mens rea* and for the development of a concept of responsibility that gives proper place to status elements. And throughout the paper, I shall use—either as a guide or a counter—Hart's concise statement of the essentially voluntary character of moral responsibility in ascertaining the role and scope of voluntary elements of awareness, intention, choice, and control in human action. Although some of my disagreements with Hart are fundamental, most involve differences of emphasis or degree, and all have developed out of periods of informative struggle with his ideas.

I

In common-law countries it is taken for granted that no man should be treated as a criminal or convicted of a crime unless he has done something wrong and knew what he was doing. Generally speaking, the behavior of a man is to be treated as criminal conduct only if there is a concurrence of *mens rea*, the awareness of the wrongfulness or unlawfulness of the conduct, and *actus reus*, the physical manifestation of *mens rea*.[2] With rare exceptions, no act or occurrence can be criminal unless the basic functions of intelligence and volition are present. The consensus of legal thinking on this point is well summarized by Professor Herbert Packer:

To punish conduct without reference to the actor's state of mind is both inefficacious and unjust. It is inefficacious because conduct unaccompanied by an awareness of the factors making it criminal does not mark the actor as one who needs to be subjected to punishment in order to deter him or others from behaving similarly in the future, nor does it single him out as a socially dangerous individual who needs to be incapacitated or reformed. It is unjust because the actor is subjected to the stigma of criminal conviction without being morally blameworthy.[3]

Generally speaking, criminal sanctions are to be applied only in those situations in which moral blame would be appropriate. If criminal punishment is understood, in the words of Professor H. M. Hart, as 'a formal and solemn pronouncement of the moral condemnation of the community,'[4] then we can more easily understand and accept the moral indignation and high-

166

pitched rhetoric of the district attorney as an expression of the lawyer's professional sense of the dependence of criminal law on morality. And unless morality makes provision for status offenses and status responsibility, we should find in criminal law a tendency to deny legal force to status crimes or status responsibility.

Morality, as interpreted by most contemporary philosophers, makes no such provisions. H. L. A. Hart, in complete agreement with Kant, lists among the distinctive features of morality the 'voluntary character of moral offenses'.[5] In developing this point, Hart writes:

If a person whose action, judged *ab extra*, has offended against moral rules or principles, succeeds in establishing that he did this unintentionally and in spite of every precaution that it was possible for him to take, he is excused from moral responsibility, and to blame him in these circumstances would itself be considered morally objectionable. Moral blame is therefore excluded because he has done all that he can do. . . . [I]n morals 'I could not help it' is always an excuse, and moral obligation would be altogether different from what it is if the moral 'ought' did not in this sense imply 'can'.[6]

In due course I shall examine Hart's statement in detail. For the moment, however, I enter it in the record as a clear statement of the generally accepted view of the essentially voluntary character of moral responsibility—a character which makes lawyers properly reluctant to apply criminal sanctions in a situation unless there is an action that is voluntary and intentional. Lawyers would be prone, moreover, to excuse any action that is unintentional, undertaken with due care and precaution, or unavoidable.

This is illustrated by the decision of the Supreme Court in *Robinson* vs. *California*.[7] Robinson was convicted under a California statute which made narcotics addiction a criminal offense. Under the terms of the statute:[8] 'No person shall use, or be under the influence of, or be addicted to the use of narcotics, excepting when administered by or under the direction of a person licensed by the State to prescribe and administer narcotics. It shall be the burden of the defense to show that it comes within the exception.' During Robinson's trial, the judge instructed the jury that this statute made it an offense for a person 'either to use, or to be addicted to the use of narcotics.'

167

The judge said further: 'That portion of the statute referring to the "use" of narcotics is based upon the "act" of using. That portion of the statute referring to "addicted to the use" of narcotics is based on a condition or status. They are not identical. . . .'[9] It was therefore unclear to the Supreme Court whether the jury found Robinson guilty, of the *act* or merely the *status* condemned by the statute. In delivering the opinion of the court, Justice Stewart complained that the statute is not one 'which punishes a person for the use of narcotics, for their purchase, sale or possession, or for anti-social or disorderly behavior resulting from their administration.' Rather, he said: 'We deal with a statute which makes the "status" of narcotic addiction a criminal offense. . . . California has said that a person can be continuously guilty of this offense whether or not he has ever used or possessed any narcotics within the State, and whether or not he has been guilty of any antisocial behavior there.'[10]

In order to categorize this statute properly, Justice Stewart considered whether it would be possible to 'make it a criminal offense for a person to be mentally ill, or a leper, or to be afflicted with a venereal disease. . . . [I]n light of contemporary human knowledge,' he concluded, 'a law which made a criminal offense of such a disease would doubtless be universally thought to be an infliction of cruel and unusual punishment. . . .'[11] He concluded that the statute under which Robinson was convicted fell into this same category because drug addiction is an illness, and that a law which 'imprisons a person thus afflicted as a criminal, even though he has never touched any narcotic drug within the State or been guilty of any irregular behavior there, inflicts a cruel and unusual punishment. . . .' The length of imprisonment was held to be irrelevant. Justice Stewart declared: 'Even one day in prison would be a cruel and unusual punishment for the "crime" of having a common cold.'[12]

Justice Douglas, concurring, compared narcotic addiction to insanity. He noted that while insane people 'may be confined either for treatment or for the protection of society, they are not branded as criminals.'[13] And he concluded:

I do not see how under our system *being an addict* can be punished as a crime. If addicts can be punished for their addiction, then the insane can also be punished for their insanity. Each has a disease and each must be treated as a sick person. . . . He [the addict] may, of course be confined for treatment for the protection of society. Cruel and unusual punishment results not from the confinement, but from convicting an addict of a crime. . . . A prosecution for addiction, with its resulting stigma and irreparable damage to the good name of the accused cannot be justified as a means of protecting Society, where a civil commitment would do as well.[14]

The opinions in *Robinson* vs. *California* thus illustrates the abhorrence lawyers generally feel for status responsibility in criminal law and their refusal to use this concept in defining crime. The opinion illustrates their confidence in and approval of the dominant concept of moral obligation which allows only for voluntary responsibility according to which only action or conduct (*actus reus*) that is voluntary and intentional (involving *mens rea*) can be morally blameworthy. Conditions of moral blameworthiness as defined by the voluntary conception of responsibility are accepted in *Robinson* vs. *California* as limiting conditions for the application of criminal sanctions. The moral inappropriateness, indeed the immorality of status responsibility, of blaming one for what he *is* rather than for what he *does*, indeed, the immorality of status responsibility is seen to be the driving force behind the Court's decision.[15]

It might seem, however, that a serious and extensive reliance on the concept of status responsibility in criminal law is found in those cases in which a person may be convicted of a crime and suffer criminal sanctions on the basis of strict liability. Many laws concerned with public welfare—e.g. laws pertaining to food adulteration or mislabeling of drugs—permit criminal conviction in cases where there is complete absence of *mens rea* and even absence of *actus reus* by the accused.[16] But even this is not a serious exception to my general point about the criminal law's avoidance of status responsibility, for the criminal law is nowhere under more vigorous or sustained attack both from within, in actual litigation, and from without, in legal scholarship, than in its reliance on strict liability.[17]

Even so it must be acknowledged that there is an increasing reliance on strict liability in tort law, particularly in cases involving the determination of responsibility for defective products, and this development has not brought shrill objection

169

from legal scholars. But here again I find no more than a highly qualified exception to my descriptive point about the rejection of status responsibility in criminal law. In the first place, tort law is a branch of civil, not criminal law: the defendant in a tort action is neither *indicted* by a grand jury of fellow citizens nor *accused* of a *crime* and *prosecuted* by a public official. And most important, he is not subject to a *criminal* sanction of fine or imprisonment; a judgment against him does not *ipso facto* imply an assumption of his moral blameworthiness or the stigma of criminal conviction.[18] In the second place, the courts have justified their reliance on strict liability in tort actions either on the principle of over-riding public welfare, or on grounds which reveal varying degrees of personal responsibility by tortfeasors even though, admittedly, the degrees would be insufficient to sustain the ascription of personal responsibility on the basis of the voluntary conception of responsibility.

To illustrate, in *Suvada* vs. *White Motor Company*,[19] we find the court saying:

Recognizing that public policy is the primary factor for imposing strict liability on the seller and manufacturer of food in favor of the injured consumer, we come to the crucial question in this case, namely, is there any reason for imposing strict liability in food cases and liability based on negligence in cases involving products other than food. . . . Without extended discussion, it seems obvious [a] that public interest in human life and health, [b] the invitations and solicitations to purchase the product and [c] the justice of imposing the loss on the one creating the risk and reaping the profit are present and as compelling in cases involving motor vehicles and other products, where their defective condition makes them unreasonably dangerous to the user, as they are in food cases.[20]

In this opinion there is no suggestion that moral blameworthiness attaches to the defendant. The court speaks of 'imposing the loss' not of 'imposing the blame' or 'imposing the penalty.' In [a] the court stresses merely the important public concern that products which are advertised and sold be safe for public use. In [b] and [c] the court points out that the party who creates the risk (whether by manufacture, advertising, sale, etc.) and reaps the profit is *in justice* the one to bear whatever loss may be incurred if such products are defective. Here the court recognizes responsibility for a loss even though

170

the loss is not the consequence of a voluntary, intentional act. The manufacturer or seller of a defective product is responsible for the loss, not because voluntarily or through negligence he occasioned the loss, but because he is the one who shaped the situation in which the loss might occur. The *status* of the manufacturer or seller rather than a specific *act* of his provides the basis of his responsibility; and yet the manufacturer or seller creates his status through prior acts even though those acts have nothing directly to do with the loss. This is, then, a model of responsibility which does not fit the models of either voluntary or status responsibility but rather suggests a conception of responsibility containing elements of both. This model should commend itself to the attention of philosophers and criminal lawyers for having at least some of the subtlety and complexity characteristic of human life.

It will be seen, I believe, that this conception of responsibility in tort law is appropriate for the concept of human action which I wish to support. But it cannot be regarded as a qualification of my basic point that in both criminal law and contemporary ethics there has been, generally, a rejection of status responsibility and an assumption of the validity and adequacy of the voluntary conception.

II

Thus far we have merely observed the rejection of the concept of status responsibility in criminal law, as represented by the Supreme Court, and in morality, as articulated by H. L. A. Hart. But whether this concept has theoretical or practical advantages of its own has not been considered; nor has the conceptual or practical adequacy in law and morality of the prevailing concept of voluntary responsibility been critically assessed. So far we have accepted passively the view of the many that the concept of status responsibility—because it conflicts with the voluntary conception—is immoral and should be rejected. But Plato has warned us about the views of the many.

If we probe beneath the surface we shall discover, I think, (1) that moral and criminal offenses cannot be understood either as voluntary actions devoid of status or as states of being devoid of intentional activity; (2) that neither status nor volun-

171

tary responsibility is adequate in law or in ethics; that both are high abstractions defying reasonable application in either field; (3) that human action, which cannot possibly be understood either as pure status (being) or pure voluntary intentionality (doing), is the complex, active being of living persons who function at various points on a continuum of action—a continuum that approaches vanishing points at the opposite extremes of pure being and pure doing; (4) that the continuum of human action is divisible into, or can be ordered in terms of, actions of distinctive types whose properties are functions of the proportion of status and voluntary elements; (5) that a sound concept of responsibility must be so fitted to the continuum of action over which it applies that it can designate the modes of response available to and/or obligatory for persons functioning at any particular point on that continuum.

Nothing less than an entire theory of human action and responsibility could establish all these points. In the rest of this paper I shall confine myself (a) to showing some of the perplexities that arise when one tries to understand certain moral experiences in terms either of the doctrine of the Supreme Court in *Robinson* vs. *California*, or that of Hart; (b) to the partial analysis of Hart's paradigmatic statement of the character of moral offenses and moral obligation; and (c) to suggesting the value and power both for law and ethics of a concept of responsibility containing both status and voluntary elements.

A

If we take a more careful and critical look at the work of the Supreme Court in *Robinson* vs. *California*, we find apparent in the court's decision the practical absurdity of trying to separate the being and the doing of human agents—an effort required by the distinction between status and voluntary responsibility. The Court agreed that the California statute under which Robinson was convicted would have been valid had it required proof of the actual use of narcotics within the state's jurisdiction. Yet if the Court was correct in defining narcotics addiction as a disease, and if one characteristic of this disease is the

172

compulsive use of narcotics, it is difficult to understand how a statute inflicts a cruel and unusual punishment by holding one responsible for having the disease, whereas the punishment would not be cruel or unusual if it were applied only to those acts which are the inevitable consequences of the disease. To use the example of the court, if 'even one day in prison would be a cruel and unusual punishment for the "crime" of having a common cold,' why would the punishment be any less cruel or unusual if it were for the 'crime' of having sneezed, coughed, or blown one's nose?

The Court was obviously on absurd ground philosophically when it tried to separate acts of addiction from the status of addiction. If the Court was right in holding that addiction implies use and that use is criminal, how could it deny that addiction is criminal? But if addiction is not criminal, then it would seem to follow logically either that addiction does not imply use or that use is not criminal.[21]

The Court could have avoided this absurdity by boldly asserting that because the status or condition of addiction and the use of narcotics are inseparable, the condition-and-use together are the disease, and that therefore neither the state of addiction nor the use of narcotics can be punished as a crime.[22] The Court was reluctant to take this step for many reasons. Paramount among them is the fact that the Court would have had to blur the distinction between the condition or state of being of the accused and the actions of the accused. This blurring would, in turn, destroy the traditionally accepted 'factual' basis for the distinction between voluntary and status responsibility.

The refusal of the Court to take this step is not entirely regrettable. There may be important uses for the fiction of pure *actus reus* and voluntary responsibility, despite their philosophical limitations. Indeed, it is one of the beauties of the law— sufficient perhaps to revive the lost faith in a Divine Order— that the Supreme Court can serve the interests of philosophers while making serious philosophical mistakes. However impossible it may be to separate the use of narcotics from the status of addiction, it is quite clear that we do not want the police arresting citizens in the absence of any anti-social behavior. Political liberty was served by the decision in *Robinson*, despite the fact that the decision reveals the artificiality and absurdity of sharply distinguishing action from being, or voluntary from status responsibility.

B

If we turn to an examination of the philosophical doctrine of voluntary responsibility as presented in Hart's paradigmatic statement, there are many points to be considered. One remarkable feature of Hart's characterization of moral offenses and moral responsibility is that he restricts himself to a pejorative context, to a context of moral failure. Hart's characterization of the distinctive features of morality in terms of moral offenses might be explained by the fact that the description is given in the context of a discussion of the similarities and differences of law and morality. There is very little in the rules of law or ethics concerning obligations to praise others or the right to claim praise for ourselves. The rules of law and morality derive their importance from the fact that they are so often transgressed; hence, most of the thought and ingenuity expended in these fields has of necessity been directed to the recognition, evaluation, and just handling of transgressions. We no more need to praise the morally virtuous than to pin medals on those who have kept out of jail. The norms of ethics and law define a high level of expectation. In law they are occasionally exceeded, but only in those moral systems providing for supererogation is there even a logical possibility of exceeding ethical norms.

But these considerations do not, in my opinion, account adequately for the negative character of Hart's exposition. Hart's selection of the pejorative context derives, I believe, both from his view of morality and from his idiosyncratic, if not dogmatic, linguistic restriction of the term 'responsibility' to situations of failure.

Hart's view of morality is essentially rule oriented. He discusses moral offense only as action that offends against moral rules or principles. If the possibility of moral offense or moral achievement is restricted to the compliance with or transgression of moral rules or principles without regard to the fulfillment or loss of moral values, moral achievement will at best be the neutral absence of moral offense. In this particular

174

context we cannot expect Hart to say everything about ethics, and perhaps he would wish to supplement his account of morality with a discussion of values. But recent English and American ethical thought has been so dominated by the discussion of rules and their many kinds that I doubt it. This is rather a bias in contemporary English and American ethical discussion that needs correction by a channel crossing and an extended vacation on the Continent.

Hart's restriction of 'responsibility' to contexts of failure would preclude, for example, our substitution of a context of moral achievement and compliance for his context of moral failure and offense, although there is nothing about the English language (or any other language) that prevents it. Suppose we use most of Hart's own words to describe an action which has not offended but has accorded perfectly with moral rules or principles. Consider the following:

If a person whose action, judged *ab extra*, has been *in complete accord with* [has offended against] moral rules or principles, succeeds in establishing that he did this unintentionally and in spite of every precaution that it was possible for him to take, he is *denied* [excused from] moral responsibility, and to *praise* [blame] him in these circumstances would itself be considered morally objectionable. Moral *praise* [blame] is therefore excluded because he has *had too little or nothing to do with it* [done all that he could do].

We may continue:

[I]n morals '*I didn't really do anything*' ['I could not help it'] is always *a reasonable disclaimer* [an excuse], and moral obligation would be different from what it is if the moral *achievement* ['ought'] did not in this sense imply *performance* ['can'][23].

Hart would surely object to the substitutions; he would never countenance my speaking of responsibility for a morally or legally exemplary act. Yet there is nothing odd about this usage. A morally good person would immediately object to being credited with responsibility for an apparently exemplary act which he performed either inadvertently or not at all; he might likewise feel some disappointment if another person were credited with responsibility for an exemplary act which he had in fact performed and for which he was in fact responsible. This is perfectly intelligible talk, and not unheard of. We also

175

find the honorific use of 'responsibility' in such statements as the following: 'He is a thoroughly responsible person,' or 'he was responsible for saving the child's life,' or 'he deserves no credit since he was not responsible for it.' There is nothing odd about these statements, but let us suppose there were. What has one proved if he establishes that a given usage is odd beside its oddness? Surely 'odd' does not imply 'wrong' or 'mistaken.' Nor could the fact of linguistic oddness, if it were a fact, offset the most important non-verbal fact about responsibility, namely, that personal involvement provides the basis for responsibility whether in contexts of success or failure. Hence we have the right to use the word 'responsibility' when there is personal involvement whether it be praiseworthy or blameworthy.[24]

With these preliminary observations out of the way, let us now consider Hart's discussion of the voluntary character of moral offenses point by point, beginning with the first sentence:

If a person whose action, judged *ab extra,* has offended against moral rules or principles, succeeds in establishing that he did this unintentionally and in spite of every precaution that it was possible for him to take, he is excused from moral responsibility. . . .

I find two ambiguities in this sentence which are not apparent or troublesome if one accepts dominant contemporary views about intention and action. For example, does 'did this' refer to an action performed unintentionally by a person, or does Hart hold that all action involves intention? If he holds that in every action the agent must have an intention, it follows that 'did this' cannot refer to an *action* but only to an event which, judged *ab extra,* might have appeared to be an action. I assume that Hart accepts the dominant view that intention is an essential ingredient in action; hence that there could be no action which was not intended. But whether Hart takes the broad or the narrow view of action—i.e., whether 'did this' refers to an action or merely to an event—is of no great importance in this context, because Hart clearly insists that moral offenses are voluntary and that actions or events (whichever word is appropriate) neither intended nor the result of negligence are not voluntary and therefore are excusable. Consequently the undetermined scope of the term 'action' results only in an unimportant vagueness so far as this passage is concerned.

176

There is a vagueness or ambiguity in the word 'intention,' however, which is of critical importance. Does Hart restrict the meaning of intention to that which is consciously intended, or would he accept the view that there are subconscious, unconscious, and organic modes of intention in addition to the conscious modes in their varying degrees of focus and intensity? By means of an examination of this passage alone there is no way to determine which alternative Hart accepts. It can be seen, nevertheless, that these alternatives confront Hart as a dilemma: the consequences of either option are inimical to his position and support the view of action and responsibility which I wish to urge. If Hart accepts the narrow conception of intention, he must sacrifice factual support for his position; if he accepts the broader conception of intention, he must blur the distinction between voluntary and involuntary to the point that moral offenses cannot be accurately designated by their distinctive voluntary character.

If we take the latter alternative and recognize varying degrees and kinds of intentionality, we recognize our personal involvement to varying degrees in complex series of events. This recognition involves our acceptance of the sequence as *our action* even though we may not have fully or even consciously intended it. On this view, we preserve the essentially intentional[25] character of all action while recognizing the degrees of action corresponding to the degrees and kinds of intentionality and personal involvement. This position, which I take, is more adequately supported by the few relevant facts available than the former alternative, which restricts intention to consciousness. But Hart cannot approve this latter alternative along with its factual support without destroying his thesis that moral offenses are essentially voluntary. For if we admit that personal involvement in action need not be accompanied by conscious intent in order for the action to be morally imputed to the person as agent, we destroy the basis for any sharp distinction between that which is and is not voluntary, thereby destroying the foundation for any sharp distinction between voluntary and status responsibility, and we alter radically the conditions or criteria of moral excusability. It is, then, clearly impossible to speak accurately or precisely of the essentially voluntary character of moral offenses.[26]

177

I presume, therefore, that Hart accepts the former alternative and restricts intention to that which is consciously intended. He holds, I believe, that a person succeeds in establishing that he did X unintentionally, if he can show that he did not consciously intend to do X. But what justification can Hart offer for restricting the meaning of intention to conscious intention— for assuming, that is, that there is no such thing as subconscious or unconscious intention? Perhaps he would rely on Stuart Hampshire's argument that: 'The sleeping and unconscious man is not an agent.... It is a necessary truth that he has no intention under these conditions.'[27] It would seem that Freud's demonstration of censorship in dreams and, indeed, the manifest content of dreams apart from any Freudian interpretation, force us to recognize the intentions of the dreamer; his personal involvement seems to color everything. Yet, like many American and English philosophers who ignore Kant's adage that concepts without percepts are empty, Hart perhaps assumes that the logic of language will supply our want of information. By restricting the meaning of intention to conscious intention and the meaning of voluntary action to that which is done intentionally, Hart can preserve the sharp distinction between voluntary and involuntary and, thereby, the basis for his insistence on the voluntary character of moral offenses. His theory gains clarity, precision, and some coherence by this move to linguistic rationalism, but it loses its factual support and plausibility.

I do not reject Hart and Hampshire's restriction of the meaning of intention to conscious intention on the basis of my intuition of the 'logic' of the term 'intention' or on the basis of my 'right' to replace their definition with one of my own. I urge, rather, that there are relevant facts about human action which are denied when one insists that a person has to be conscious of what he intends in order to have an intention. Just as the anatomy and organic functions of the whale force us to admit that a whale is not a fish, regardless of the logic of the term 'fish' or the definitions of venerable dictionaries, the anatomy and dynamics of human behavior and action force us to recognize unconscious and subconscious, no less than conscious, intentions.

Factual support for the broader conception of intention, and for the theoretical implications regarding responsibility that follow from it, is found in abundance in the daily affairs of ourselves, other individuals, and nations. How am I to regard those movements of mine which are judged by others, *ab extra*, to be my actions and which may reveal to others one or more of my overriding, long-range intentions, but which I can truthfully report were not a part of my conscious intention at the time my movements took place? Consider the way, for example, men and nations pick fights and exacerbate quarrels to their enormous advantage while truthfully and conscientiously denying all conscious intent or desire to fight. Consider Oedipus' attack on Creon; did he really believe that Creon was guilty or was this just the sort of conduct to which Oedipus was habituated? Consider Odysseus' trifling, yet possibly sincere, excuses for failing to support Hecuba in Euripides' play; what did he really intend? Consider the actions and intentions of Hitler and Chamberlain prior to the outbreak of World War II. In these cases we have factual proof of action possessing an intentional structure, or a goal-direction, radically at odds with its conscious intention.

And such actions must be judged morally. We are prepared, I believe, to say that both Hitler and Chamberlain were morally blameworthy men, the one for his almost diabolical craving for universal destruction (however much he may have spoken of peace and German fulfillment)[28] and the other for his cowardice and preoccupation with immediate selfish advantage (however much he may have spoken of reasonable compromise in the interest of peace).[29] Are we not prepared to recognize, moreover, the corporate moral guilt of the Englishmen who cheered Chamberlain on his return from Munich and of the Americans who relied on Washington's Farewell Address to justify avoidance of entanglements on behalf of freedom in Europe? But did these Englishmen and Americans consciously intend to behave as cowards or to evade their moral obligations on the Continent?

More prosaically and perhaps more convincingly still, consider the Christmas dinner at which the spinster aunt, in the shrill voice of Carry Nation, delivers a temperance lecture while the father is opening a bottle of wine saved for the occasion. Are we to deny that the aunt's dislike of the father for having destroyed her only immediate family by marrying her sister, and the aunt's envy of her sister for being the mother in another family, are expressed in her action? Are we to believe that she does not desire and intend to hurt this family, to dampen the pleasures of its Christmas feast? Yet who would call the aunt a liar when later, in tears, she apologizes for having spoiled the celebration while she continues to insist that her only concern was for the welfare of the father and mother and children who are going to destroy their health by drinking? The aunt can claim, with complete justification, that her love for the family has been fully demonstrated by her generous and loving support in times of extreme hardship at great personal sacrifice to herself. But it is equally true that she is resentful of the family and full of hate. And since her actions are *hers*, it is not surprising that they should reveal much more of herself than she consciously intends to express: what she *does* is a function of *all* that she *is*, all her loves, hates, and wants, and not merely the expression of what she consciously intends when she acts. In this case all of her action, including its disruptive consequences and its good and bad will, was intended and was done intentionally, despite the fact that her conscious intention was merely to save the family she loved from alcoholism. The most accurate account of this situation is one which simply accepts *as fact* the presence of unconscious and often ambivalent intentions.

Five years ago in Austin, Texas, a man charged with the murder of his wife claimed in his defense that he had killed her while sleeping or immediately on awakening.[30] Psychologists and psychiatrists testified to the possibility of this occurrence. And if the man had lived happily with his wife for twenty years, one might be inclined to excuse his act on the grounds that it was unintentional. According to the evidence, however, the man and his wife fought frequently, and he had planned on two occasions to divorce his wife in order to marry another woman. By his own admission he dreamed that he was killing an intruder who was chasing his nieces before awakening to find that he had killed his wife. I would accept as factual that the man killed his wife while in an unconscious or subconscious state. But I see no reason for concluding, as Hart or Hampshire would, that the man did what he did unintentionally and, consequently, that he is to be excused from moral blame for

180

killing his wife. (Nor would I argue, on the other hand, that he should be found guilty or punished in a court of law on the basis of these facts unless and until careful safeguards and limitations have been developed for the introduction of such evidence. I fully recognize that the implementation of my theory in legal practice requires the development of solutions to a host of special problems. Neither the problems nor their solutions can be dealt with here.)

Hart insists that a person is to be excused from moral blame if he can establish 'that he did this unintentionally and in spite of every precaution it was possible for him to take.' Now if we admit that the man killed his wife while in a non-conscious state and that consciousness is required for intention, it follows that in this case the man did what he did unintentionally. And unless we consider him reckless or negligent for having continued to sleep in the same bed with his wife after having quarrelled with her,[31] we have no basis for claiming that he failed to take every precaution it was possible for him to take. We cannot fault him for having failed to consult a psychiatrist or marriage counsellor; recourse to such professional help presupposes a level of education and sophistication which the accused had not attained.[32] On Hart's view, we must conclude, therefore, that the man was neither negligent nor intentional in his behavior, hence that he did not offend voluntarily, hence that he is excused from moral responsibility and blame.[33] This, in my opinion, constitutes a *reductio ad absurdum* of the view that limits ascription of moral blame and responsibility to voluntary, consciously intended acts, the view that moral offenses must be voluntary and that in order to be voluntary they must be consciously intended.

There are at least two ways of avoiding this absurdity and acknowledging the moral responsibility of the agent in this case. First, we may hold that, since the man had no conscious intention to kill his wife, his bodily movements in killing her do not constitute an action, and hence that killing her was not voluntary. In this way we preserve the usages of Hart and Hampshire. But then we are forced to abandon Hart's thesis that moral offenses are voluntary: we are forced to predicate the man's responsibility on his being, or status, rather than on his voluntary action. We now avoid the absurdity of excusing him by morally blaming him for *being* a man who killed his wife even though he did not kill her voluntarily or intentionally.

Or, second, we may describe what the man did in terms of action, intention, and volition developed on a continuum view of responsibility. We may hold that there was a degree of voluntariness in his action proportionate to the degree and kind of intentionality and, consequently, a corresponding degree of moral blame. We would have to assess the degree and quality of his intention by reference to what he had thought, said, dreamed, and done about his wife in the preceding months and years. And to the degree that the man's intentions, as so assessed, were apparent in his bodily movements of killing her, we would describe those movements as, to that degree, his voluntary action.[34] This second way, by far the soundest in my judgment, provides a better fit of facts to theory than the first way. It imposes, moreover, very little strain on traditional linguistic usage and offers qualified support for the traditional view, represented by Hart,[35] of the voluntary character of moral offenses.

On the second view we recognize that there must be some element of intention and some degree of voluntariness in a series of bodily movements if those movements are to be called an action and if the action is to be subject to moral judgment. At the same time, moreover, we recognize the essential co-presence of elements of status responsibility. The mixture of kinds of responsibility reflects with accuracy the mixture of being and doing in personal action. It reflects the fact that what a man *does* is a function of what he, in the context of his situation, *is*, and that what he *is* within this context is revealed by what he *does*. The partial truth of the voluntary conception of responsibility is acknowledged through the recognition that what a man *does* is the *ratio cognoscendi* of what he *is*, and the partial truth of the status conception of responsibility is acknowledged through the recognition that what a man in context *is* is the *ratio essendi* of what he *does*. This view can also accommodate the existential point that what a man is and does determines or creates what he shall be and do; that his existence can give rise to a new essence.

We are compelled then, largely by factual considerations, to reject the view that intentions are necessarily or always con-

182

sciously intended, that a person must be conscious of his intention in order to act intentionally. We are forced, that is, to reject the view that a person can establish that he acted unintentionally if he can show that he did not consciously intend to do what he did. And when the concept of intention is extended, the character of moral offenses and the criteria of moral excusability are altered; we recognize, for example, the possibility of being morally blameworthy for what we do on the basis of unconscious or subconscious intentions in the absence of any conscious intention to violate moral rules or principles or to neglect any values that should be enhanced.

And we have taken only the first step toward confronting factually and acknowledging theoretically the larger range, scope, and depth of mental concepts. The enlargement of the concept of intention must be accompanied by a comparable enlargement of the concept of awareness. The importance of this step can be seen most clearly in the present context if we ask whether a person can be morally blameworthy for negligent conduct. Hart recognizes, of course, the difference between intentional or purposive action and action which is done with knowledge or conscious awareness but without intent. He insists that the person whose action violates some moral rule or principle must, in order to excuse his conduct, establish not merely that he acted unintentionally, but also that he did not know that he was running any avoidable risk of violating them. He must establish that, in Hart's words: '[H]e did this . . . in spite of every precaution that it was possible for him to take.' Thus, in order to be excused morally, the person is required to prove that his conduct was not reckless. But it is important to note that he is not required to prove that his conduct was not negligent. As exemplarily explained in the *Comments* to the Model Penal Code,

recklessness involves conscious risk creation. It resembles acting knowingly in that a state of awareness is involved but the awareness is of risk, that is of probability rather than certainty.[36]

Hart's view clearly requires that a person who would excuse his action must succeed in establishing that his conduct was neither intentional nor reckless: he must show both that it was not his purpose or intent to violate the rules and that he was not aware that he ran any avoidable risks of doing so. But on Hart's view a person is not required to show that his action was non-negligent, for negligence, unlike recklessness, does not involve any state of awareness:

It is the case where the actor creates inadvertently a risk of which he ought to be aware, considering its nature and degree, the nature and purpose of his conduct and the care that would be exercised by a reasonable person in his situation.[37]

If Hart takes a narrow view of mind, intention, and awareness, and if he holds that moral offenses must be voluntary—by which he means that they must be avoidable by means available to the agent at the time he acts—how can the agent be blameworthy for failing to take a precaution of which he was not aware and which was therefore not available to him at the time he acted? If the agent is aware of reasonable precautions which he is neglecting to take, he is acting recklessly. But if he is not aware of any reasonable precautions that he is neglecting to take, in what sense can it be possible for him to take them? In what sense can he be *voluntarily* negligent and therefore morally blameworthy for his negligence?

It makes no sense to include among possible precautions that one ought to take precautions of which one is not consciously aware unless it is recognized that there are various modes of awareness, including peripheral, subconscious and unconscious modes, and that there are purposive acts of forgetting, repressing, neglecting, etc. If we accept the fact that persons express through their actions intentions of which they are not fully or even partially conscious, and if we give credence to the psychoanalytic and psychological evidence of repression and other forms of subconscious or unconscious awareness and activity, then—but only then—have we a sound factual basis for extending the concept of moral blameworthiness to truly negligent behavior. For it is only after we accept this evidence as factual that we have a basis for identifying the presence and effects of the person in such conduct. It is another shortcoming of the traditional view that moral offenses must be voluntary in the sense that they must be avoidable by means of which the agent is aware, that it limits moral blame to actions which are either intentional or consciously reckless and, hence, that it cannot impute moral blameworthiness (or praiseworthiness) for truly negligent conduct.[38]

On the basis of the extended view of awareness and intentionality, by contrast, it is possible to hold a person morally blameworthy and legally culpable for genuinely negligent conduct. And by the extension of the meanings of these mental concepts—in response to factual evidence, be it noted, and not to the 'logic' or usage of these terms—we need not abandon but only qualify the traditional requirements that morally blameworthy acts be voluntary and that legally culpable conduct involve *mens rea*.[39] In holding a person morally blameworthy or legally culpable for truly negligent behavior we recognize the presence of the person existing and functioning mentally on some level in the process of acting. We recognize degrees of moral blameworthiness or legal culpability appropriate to the degree and kind of personal presence in the action. The greater the degree of consciousness in awareness and intention, the greater the degree of voluntariness and *mens rea*; hence the greater the degree of moral and legal responsibility.[40] Once again we confront increasing or decreasing continua of awareness, intention, *mens rea*, etc., on which our judgments—on continua scales—of the presence and degree of personality, voluntariness, action, responsibility, blameworthiness, and culpability are based.

Because there is still much doubt (a good deal of which may be fully merited) about the soundness and relevance of data provided by depth psychology and psychoanalysis, I do not want to rest my case against the traditional, simplistic concept of voluntary responsibility exclusively or even primarily on such data. By restricting ourselves to familiar experiences in daily life—without appeal to psychoanalytic interpretation—we can expose the inadequacy of the view that a person can excuse himself for morally offensive conduct by showing that he did what he did unintentionally and after taking every possible precaution. According to Hart's statement, if a person meets these criteria: 'Moral blame is therefore excluded because he has done all that he can do . . . [and] in morals "I could not help it" is always an excuse.' But we can show, I think, that 'I could not help it' is not always an excuse because moral responsibility can have no meaning in human affairs unless there are times and situations

185

in which one is morally responsible (deserving of moral praise or blame) for what he is, whether he could have helped being what he is or not. That is to say, I wish to show by reference to uncontested facts of human experience that the concept of moral responsibility (and, by limitation, moral offense) involves some minimal element of status responsibility and cannot be based solely on voluntary responsibility. To show this at least sketchily will be the burden of the final part of this paper.

III

It is often mistakenly assumed in philosophical discussions or action, intention, person, and responsibility that every one is clear about the precise, and even logical, difference between an event and an action, between an action, and its consequences, between a voluntary and an involuntary action, or movement, between a person and a thing. In fact, however, there is great uncertainty and fuzziness on all these matters: wherever we look we seem to find one item or concept fading by imperceptible degrees into another from which it is alleged to be factually or even logically distinct.

A

We find, for example, not merely the gradations of personality, agency, and responsibility in the sequential observation of comatose, vegetative, senile, idiotic, infantile, stupid, sleeping, insane, neurotic, normal, wakeful, rational, articulate, intelligent or imaginative persons, we also find a spectrum of action, personality and responsibility in the daily life of any ordinary human being. Consider the following experiences of X:

1. While walking aimlessly in his garden, he steps on a thorn and feels a terribly sharp pain.
2. While playing badminton in his garden, he steps on a thorn and feels some pain.
3. While in a desperate struggle with an intruder in his garden, he steps on a thorn and feels no pain at all.
4. While asleep he dreams of a stranger who is killed and whose estate is inherited by his brother.
5. Working in his garden while hungry he thinks suddenly of eating bacon and eggs.

6. Working in his garden he thinks of his brother who is on military duty in the war zone, and he offers silent prayer for his safety. In the midst of the prayer the thought crosses his mind that if his brother is killed, he will inherit his brother's estate.
7. Hungry but still at work in his garden, he decides to cook those eggs and bacon.
8. As he is going inside, he thinks, 'I don't want my brother dead; what a scoundrel I must be for having a thought like that.'
9. He prepares lunch.
10. His brother comes in unexpectedly on a military leave granted so that he can recuperate from a wound and lead poisoning; X invites his brother to eat with him.
11. X decides to slip a fatal dose of powdered lead into the eggs before serving his brother.
12. X puts the poison in the eggs.
13. He serves the eggs to his brother.

Here we have a continuum of situations from events to moral action in which a gradual increase of personal involvement and responsibility is shown. At what point shall we speak of action rather than mere event? At what point does the personality of X express itself in what happens or in what is done? At what point do we speak properly of moral responsibility or moral blameworthiness? Of legal responsibility or culpability?[41]

X's personality and personal involvement are apparent from the outset. Even the way in which X feels pain in 1 has elements of action about it. The intensity of his pain is a function, presumably, of his degree of abstraction while walking, and of his normal pain threshold. If he has a low threshold and vivid memory of such experiences, the pain may be excruciating and he may relive the shock for hours or days. If his threshold is high a single 'Dammit!' and the removal of the thorn may be all there is to it. Now are we to suppose, in the interest of precision or clarity, that X's reaction to stepping on a thorn is just a *reaction*, a psychophysical event in which there is no personal involvement and no element of action? Can we doubt that X's response will be not merely indicative but largely determinative of his action in a situation of moral crisis in which the threat of

187

pain is involved? If, while later serving in the army, X were taken prisoner, how would he respond to the mere threat of physical torture? Can we assess his moral responsibility by asking 'Could he help doing whatever he does?' If his pain threshold is low and his memory of past pains and his imagination of pains to come are vivid, can he help divulging secrets on the mere threat of torture, whereas he would not divulge them even after torture, were his pain threshold high and his imagination and memory less vivid? Does it make any more sense to say 'He could not help doing what he did' than to say 'He could have avoided being who he was'? Or does it make any more sense to say 'He could have avoided doing what he did' than to say 'He could have avoided being who he was'? If what a man does is not a function of what he is, in what sense can his action be his? But if what a man does is a function of what he is, such questions make no sense. The proper question for the assessment of moral responsibility should rather be: 'What kind of person is he—that is, under what conditions, both external and internal, does he do or would he do what he did?' I see no way of determining whether or not X can be different from what he is or could be different from what he was at any particular time. Likewise, I see no way to determine whether he can do differently from what he does or could do differently from what he did at any particular time. But there are ways to determine to some extent the conditions under which X does what he does and is likely to do what he will do—that is, we can come to know something about his character, including his moral character, and a statement of his character is a description of his being-doing.

Now if X screams and cries when he steps on the thorn in 1, we may be able to talk to him about his behavior and train him so that he will exercise greater control on the next occurrence. When on the next occurrence he shows greater control, stiff upper lip, etc., are we to say that praise is inappropriate for it is only another event and not an action for which X can be praised? Hardly. The fact is that we train children to exercise control—and hence to *act* to a minimal degree—even in the way that they experience pain. I can hear a critic saying: 'But you train the child to control his response to the pain, not to alter his experience of it.' It is probable that control would be impossible in cases of intense pain if there were no way of altering the experience itself. The unity of mind-body in human experience has been seriously underestimated by philosophers since Descartes. They have also underestimated or ignored the personal controls that we know are operative, but which we are not aware of as operating in such basic processes as perception.

But if one questions the presence of minimal personal involvement and therefore a minimal element of action in 1, what will one say about 2 or 3? If we argue that the adjustments in the awareness of pain made in 2 and 3 are merely bodily adjustments having nothing to do with the person involved, we shall be left with a high abstraction instead of a richly concrete person. The extent of the reduction in the awareness of pain in 2 is not merely a function of the attention areas in the brain; it is likewise a function of X's involvement in the game. If he doesn't like the game, the pain is likely to be far more intense than if he were extremely fond of it. The greater his competitive involvement, the less intense the pain. If he is playing with a young woman in whom he has a strong romantic interest, his pain may be either lessened or intensified according to his courting technique, quite apart from the question of whether he will feign greater or lesser pain for courtship purposes. His personality will express itself instantaneously in the midst of play prior to his conscious assertion of secondary control. I see no reason to deny that the initial response is his personal response, not merely an organic reaction, though I should not want to deny the greater element of personal involvement expressed through his secondary control.

I acknowledge that pain must not be too intense if the element of personal action is to be found in the very perception of it. Pain so intense as to produce almost instantaneous loss of consciousness is clearly of a different sort. But this consideration should not blind us to the minimal expression of the person in the experience of less severe pains.

In a situation like 3 we should condemn a soldier or an athlete who did not suppress virtually all awareness of pain. (I introduce athletes into a situation like 3 out of consideration of American pro-football which is much more like 3 than 2.) If he continued to feel pain from a thorn to the point that it interfered with his fighting, we should probably deny that 'He did all that he could to control the pain.' (Note we do speak of controlling pain when we can mean nothing other than controlling the way we experience it rather than the response we make to it.) We should be inclined, I think, to say that a football player who felt enough pain from a thorn in the foot to be seriously distracted by it while engaged in hand to hand 'combat' cared too little about winning the game and 'had not done all he could do' to win it. Whether or not we regard his failure as morally blameworthy depends on how seriously we take the game and whether we view it as a moral struggle, but not on the character of his action. His personal involvement, minimal though it be, is sufficient to justify imputation of some very small degree of moral blame.

By the time we come to situation 4, I should suppose the presence of personal involvement and responsibility would be generally acknowledged. It was Plato, not Freud, who first stressed the moral significance of dreams and who insisted that it was important to consider a person's dreams when assessing his moral character. It is universally acknowledged that dreams reveal the desires of the dreamer: starving men dream of sumptuous meals; sexually deprived persons dream of sexually pleasing objects; bed-wetting children dream of toilets while they wet their beds. Now if we add to these commonplace observations Freud's theory of dream censorship, we find an important similarity between 4 and 6. Freud would morally credit X for his dream work, for his censorship of his dream. He would assert that X's love or respect for his brother, or at least his acknowledgment of his brother's right to live, was expressed in his suppression of the true content of his dream and in his provision of the manifest content. Freud would say that X was a better man for dreaming what he dreamed in 4 than if he had dreamed directly of his brother's death, for his dream in 4 shows that he disapproved of his own desire.[42] Without trading on the metaphysics of psychoanalysis, I should argue that X's incompatible wants were revealed in 4: his affection for his brother is in conflict with his desire for his brother's estate. And I should argue that we can never make sense of personal or moral responsibility unless we recognize the expression of the person and hence a mode of personal action, if not in one's dreams, at least in one's wants and desires.

190

By moving from 4 to 5 we confront essentially the same issues but in a context in which the rejection of Freud's or Plato's interpretation of dreams poses no threat to my position. But I do not wish to deny what seems to me the clear moral relevance of dream behavior. I contend that no one would knowingly hire a babysitter who frequently dreamed of killing kittens or chickens, much less one who dreamed frequently of killing children. We all know perfectly well that our dreams reveal ourselves, our persons. When I was a child of eight—and long before I had heard of Freud—I dreamed on two occasions of the deaths of my parents. In these dreams I basked in the emotional glory of being the 'poor little orphan.' When I awoke I was ashamed of myself for indulging in those gratifying thoughts of having everyone sorry for me at the cost of losing my parents. I considered then, and now consider that a moral fault (admittedly of trivial importance) was revealed in those dreams. But I would not know what it means to be me or what my moral quality as a person were unless I based my judgment on all indices of myself. I might add that in dreams in which I have 'done the right thing' in a situation of great temptation I have awakened mildly pleased with myself. Only a fool would ignore what he can learn of himself through his personal activity in dreams.

For the sake of those who reject Joseph along with Plato and Freud, however, there are always 5 and 6. If in a fully conscious state I think of the attractive consequences of my brother's death in a situation in which I am concerned for his well-being (I need not be praying about it), I must obviously recognize my ambivalence toward him. And if I have no basis for wishing him ill, must I not recognize my personal involvement and agency in the morally blameworthy thought of the attractive consequences of his death? Let me emphasize once more that we are speaking here of microscopic blame. X could not be called a basically morally bad person because he had the dream in 4 or the thought in 6 unless they were interpreted later in light of 11, 12, or 13. But the continuum of personal activity and responsibility is what I wish to stress. And I think we find in 6 a significant though minuscule instance of personal agency, responsibility and, indeed, moral blameworthiness. The degree is far greater, moreover, than in situations 1 through 5.

Situations 4 through 6 show that a person's wants are a part of him and an expression of his personal agency. A man who will not assume responsibility for his desires and wants may just as well deny responsibility for his mind-body. I believe that what I say accords with the findings of clinical psychology in so far as the person who does not recognize himself in his desires and wants and in the subtle ways in which their intensity bespeaks his control of them, antecedent to consciously deliberate control, is suffering some degree of mental illness. (Ego is no longer master; repression is terribly extended, etc.)

In situation 1 through 6 we have the gradual emergence of clearly recognizable action from events in which faint but significant traces of personal activity are found. In situations 6 through 13 we have actions in which there is a gradual development of conscious deliberation and moral responsibility. Numbers 11 through 13 can be described as one action or as three, but the quality of the decision in 11 is not fully revealed until the occurrence of 13. X's decision in 11 may not have been fully determinate even in the mind of X at the time it was made. His decisiveness *becomes* complete as his action develops. But even at 13 there may be irresolution: 13 may be followed by a 14 in which X snatches the plate away before his brother can eat. The person of X may change and develop in the tension of the situations. But his being will never be separated from his doing; he is and does together and at once. It makes no sense at any point to say that he could have *done* differently unless he could have *been* different.

Of course there were alternatives open to X at every point from 1 through 13 and, as we noted, alternatives remain open after 13. But were there alternatives which X could have taken while being what he was at each instance? This is the question we cannot answer with any empirical guarantees. Only the metaphysicians of free will or determinism can fight this issue through. But we can recognize the important differences in the quality or kind of alternatives available to X. We note that his awareness and conscious deliberations are on an increase from 1 through 13. His alternative at 1 may be little more than the possibility of accepting or rejecting himself as the person he is and deciding on a course of training to raise his pain threshold. Indeed in all situations from 1 through 5 his action or activity is not so much planned as happening. And his act of praying in 6 is a sudden impulse (though a morally significant and revealing one). But in 7 there is planning. He decides on a course of action—which is itself an action or part of one—before he goes inside or gathers the food and the utensils and begins to cook. Time passes between the action of 7 and the cooking of the eggs. And in every moment of time there are occasions in which his person can express itself in different ways if his person is such that the expression comes forth. But time, the *sine qua non* for alteration of plans, for the expression of ambivalence or contrary desires or intentions, for deliberation and thoughtful consideration, is provided. And the person whose action has been undertaken and sustained through prolonged moments of time, in which deliberation and consideration of alternative desires, wants, intentions, and plans may express themselves through the alteration of the course of that action, is more fully identified and identifiable with that action than a person whose action is of less duration down to the point of being almost instantaneous. Enduring action in which the full capacities of the person are engaged is what is properly meant by voluntary action: it is expressive of the volition of the person; it reveals what the personal agent wants to come to pass in the world and what he wants to be.

Concerning voluntary action, we can say that there was time and opportunity for the agent to do differently from what he did had he been a different person from what he was. Hence we are prepared to blame him more for such action, because his personal identification with and in such action is greater. In blaming him severely only for voluntary actions, however, we are not denying that we are blaming him for what he *is* as well as for what he *does*. We blame him more severely because his fully conscious mind and deliberative choice are expressed in temporally extended voluntary action. We blame him less severely for impulsive or responsive actions or for dream actions precisely because far less of himself is expressed in or identified with them.

B

In section A we examined an event-action continuum in which we observed the gradual increment of personal expression as we moved from cases of predominantly event-like reaction, through those involving sudden responses, dream work, impulsive thoughts or desires, to cases of maximally conscious, voluntary and deliberative action. The continuum was one of increasing personal involvement and expression, increasing voluntariness, and increasing responsibility. Throughout we found, moreover, the co-presence of personal being and personal doing, and we observed the artificiality that results from the separation of the person's being from his doing.

In section A we observed and considered only a small aspect of the ontological foundation of action and responsibility. Action and responsibility depend on far more than the being and doing of the agent himself; they depend also on the being and doing of other agents and finally upon the general matrix of action, including all of the ontological conditions on which action depends. The ontological matrix of action no less than the intentions of the agent sets the determinate limits of action and the degrees and quality of responsibility.

Let us resume our consideration of X by supposing that each of the following situations is an alternative successor to 13:

14. X snatches away the eggs just before his brother eats them.
15. X's brother begins to eat, feels sick almost at once and stops eating before consuming a fatal amount; X throws away the eggs (*a*) in happy relief that the plan has failed, or (*b*) in anger that his plan has failed.
16. X's brother eats the fatal meal and dies.
17. After serving the poisoned eggs, X leaves the house for a few minutes; while he is gone his mother enters, partakes of the poisoned meal, and (*a*) both she and the brother die, or (*b*) the brother feels sick, does not eat, and only the mother dies.

In 14, X is still in control as much as any voluntary agent is in control of his action; that is, his action has not yet set in motion or been caught up by forces that may result in a disrelation between his action and his plan or intention. In 14 we see that the determination, expressed with increasing clarity and force from 11 through 13 is still far from steadfast or overpowering. In 14 we find that determination is shaken by competing aspects of X's personality and interests; his action now expresses per-

194

haps either his prudential concern for his own safety and well-being, his continuing but ambivalent love for his brother, or his respect for law. X's movement in 13 may have been impulsive: an expression on his brother's face may have reminded X of a happy incident from their childhood. Alternatively, his movement may have been deliberate: he may have known, even as he was serving the plate and carrying it to the table, that he would have to snatch it away at the last; perhaps he only toyed with the idea of murder and even savored the moral test he was putting himself through. The range of possibilities is almost infinite.

But 14, whether impulsive or deliberate, forces us to reassess X's blame. Our judgment made on the basis of 11 through 13 is no longer adequate. If X's action from 11 through 13 was deliberate while his action in 14 was impulsive, can we allow our judgment based on 11 through 13 to stand? Shall we argue that 11 through 13 prove that he intended to murder his brother and that it is immaterial that he was stopped by an expression on his brother's face rather than, as in 15, by the fact that his brother does not eat the poisoned food? Shall we argue that he was stopped in either case by an accident in so far as he was concerned? We must argue, I think, that he is to be credited morally for his response to his brother's expression in a way that he is not to be credited for his brother's refusal to eat the food—though both accidents, if that is what they are, partially determine his moral worth. Much of X is centrally involved in his response to his brother's facial expression: it is not just the expression but the expression as seen by X that accounts for X's throwing away the eggs. The expression as seen by X shatters his resolution and alters his intention, whereas there is no break in X's intention in 15(b). We must give X moral credit for his response to his brother's expression, or for his deliberate consideration in which his love for his brother was reasserted, or for his thought of his brother's right to live and the wrongness of murder, no matter how intensely he may have hated his brother.

We must give X moral credit, though to a lesser extent, even if his action in 14 expresses nothing more than his prudent concern for his own safety; there is an important element of moral goodness in the man who is law-abiding even for the wrong reasons—he does not destroy some of the values that the laws protect. Or are we so carried away by the moral dare-deviltry of formalistic, voluntaristic ethics that we find no moral worth in law-abiding conduct which is prompted by selfish motives?

The situation in 15 seems radically different from 14, but it is only gradationally different. In 15 the action has clearly moved beyond the voluntary control of X. But did X have voluntary control over his brother's expression in 14, or over the present strength of his love for his brother, or over the education that developed whatever sense of duty and respect for the lives of others that may have moved him in 14? We know what X was in 14, but we must not slip back into the mistake of supposing that X could have *done* differently in 14 any more than in 15 without having *been* a different man in these situations.

Since the situation in 15 is none the less beyond the voluntary control of X, what should we conclude concerning his responsibility? His maximally voluntary responsibility must be assessed by reference to his reaction to the failure of his plan—to the truncation of his action. In 15(*b*) we find that his resolution was complete: X did his best to kill his brother and never wavered in his intention. We can say, as I suggested in another paper,[43] that X murdered unsuccessfully. But that is a misleading way of putting it, for it is a brute fact that in 15, X did not murder his brother. The continued existence of his brother provides an ontological refutation of any charge of murder. X has been saved by his brother's sensitive digestive system from the crime of murder. Although X did all he could to bring it about, although his voluntary involvement was complete, his action was terminated short of its completion. His action was defined by his intention to kill his brother, but his action was terminated (*a*) after he had done all that he could to realize his intention but (*b*) before his action had been completed by his brother's death. In 15(*b*) we have a situation in which the ontological matrix of the action does not support the volitional matrix, and the discrepancy is effectively articulated, in my opinion, by the traditional language of volition. X *willed* to kill his brother and is morally blameworthy for his acts of volition even though he did not succeed.

But is X as morally blameworthy for having willed, but failed, to kill his brother, as he would have been if his action had fitted

196

his intention; if, that is, his volitional act had been completed by his brother's death?[44] Unless we banish from ethics all concern for the realization of values, our answer must be that X is not so blameworthy. In spite of his volition X is not guilty in 15 of having destroyed a human life with all its values! If in 16, X can be held responsible morally for his brother's action in eating the eggs and for the action of the poison within his brother's system, why should he not be relieved of some responsibility and blame if these intended and probable consequences of his efforts in preparing and serving the eggs do not take place?

The traditional answer has been that X intended the consequences and did everything in his power to bring them about; hence, they are a part of his action and he is morally accountable for them whether they happened or not. Persuasive as this answer is, it overstates the case. It ignores the absence of certain elements of being or status requisite to full moral responsibility. The man who has attempted murder, as X has in 15(*b*), is as guilty *volitionally* as he can possibly be. But he lacks the *being* or *status* of a murderer. It would be a *reductio ad absurdum* of the theory of voluntary responsibility to assume that he would not be far more blameworthy had he acted in an ontological matrix that supported his intent and brought about its full realization. On the other hand, it would be a *reductio ad absurdum* of the theory of status responsibility to assume that the man who has attempted but failed to commit murder has no moral guilt as a murderer just because his victim is still alive and unharmed. His volitional offense still stands.

Much more needs to be said about the ontological matrix of action. But it is already clear that it can alter the agent's action despite his intention or volition and, hence, that it can alter his moral and legal responsibility. If we limit our considerations merely to X's awareness of himself and ignore what the law or his family might think of him we must recognize the difference in what he as a person is in 15 (whether (*a*) or (*b*)), and what he is in 16. According to 15, X is not a murderer although he has a murderous will in 15(*b*) and may have no better than an ambivalently murderous will in 15(*a*). In both 15(*a*) and 15(*b*), however, there are redemptive possibilities open to X that are closed by 16. Reconciliation with his brother is only the most obvious. By considering the difference between X in 15 and X in

16 we see plainly his finitude both physically and morally: he is dependent with regard to both his moral guilt and virtue on many ontological factors that are not under his control, not even in the weakened sense of being expressions of his volitional being. This fact of the dependency of his moral virtue or guilt on ontological factors beyond his control is not acknowledged by the traditional view of moral responsibility according to which one deserves credit or blame only for what he has done voluntarily. The traditional view necessarily ignores this fact because it has ignored the ontological foundations of voluntary action.

In 15(a) we have a situation that is similar in many respects to one of the situations we considered in interpreting 14. X's resolution is still divided and incomplete. He is still ambivalent; in many ways he is still the man he was in 6 and 8. *Fortunately*[45] he is not a murderer with regrets and remorse but only a man who has come very close to being one. He is protected from or relieved of some moral blame by the collapse of the ontological matrix required for the completion of his murderous intent. X's moral blameworthiness under 15(a) must, nevertheless, be substantially greater than it is under 14, no matter how we construe X's motives in 14. There is an incremental rise in X's blameworthiness as we move from the interpretation of 14 as motivated by X's respect for law to 14 as motivated by prudence. In 14 *his* being and doing are largely decisive; in 15, however, his intended action is truncated by the collapse of the larger ontological matrix for which X has far less responsibility. Since X has far less to do with the truncation of his action in 15 than with his own termination of his action in 14, there is no basis for reducing his moral blameworthiness in 15(b) at all, or in 15(a) more than slightly, in so far as it is based upon elements of voluntary responsibility. But X's blameworthiness based on elements of status responsibility is not substantially greater in 15 than in 14, though it rises sharply in 16.

In 16 X's act is fulfilled, completely realized. His intention and volition are fully and accurately expressed in and supported by the ontological matrix which includes his brother's act of eating and digesting and the poison's causal efficacy. X would not and could not be a murderer without the support of this or some other matrix over which he has no control. In 16 there is no increment of intention, determination, or voluntariness in

X's action over what was present in 15(*b*); the increase in his moral blameworthiness in 16 over 15(*b*) must come therefore from an increase in his status or ontological responsibility. The fit of intention and volition to the ontological matrix is perfect: the full action is expressed in this absence of disrelation between intention and occurrence, between volition and being. The full action is morally imputed to X because the person of X is so transparently present in this fusion of doing and being.

Another sort of disrelationship between action and intention is introduced by 17 (for the sake of brevity we will concentrate on 17(*b*)). Here we observe the extension of an action beyond the limits intended by and directly influenced by the agent. We may suppose that X had no relatives or friends in the city other than his mother and brother; we may likewise suppose that neither the mother nor the brother had friends in the city; and finally, we may suppose that X put his mother on a train to another city far distant from his own on the very day the action took place. Now X's brother cannot be blamed for sharing with his mother a meal that he believed to be wholesome: he was neither purposive, knowing, reckless, nor negligent in feeding her the poisoned eggs. But what shall we say about X?

We may assume that X was intensely (though quite properly and non-Freudianly) fond of his mother; we may even trace part of X's hostility toward his brother to their competition for her affection. Under these quite reasonable suppositions we see that X is as free from moral blame for the death of his mother as his brother is—if we hold to the view that moral offenses are voluntary. X did not intend to kill his mother any more than his brother did. X did not know that his mother was in even the slightest danger of being killed. Far from being negligent or reckless, he took the precaution of recalling that he had put her on the train to another city and watched the train pull out only a couple of hours before his brother arrived. X was not purposive, knowing, reckless, or negligent in so far as the poisoning of his mother was concerned. It makes no sense to say that killing his mother was his voluntary act.

At this point we must be clear about two facts: first, we know that X is guilty of crime in 17(*b*); second, we know that he is morally blameworthy in 17(*b*). These facts are not in doubt.

The problem is: What are the essential characteristics of a theory of moral responsibility that can account for these facts?

In law and morals the problem has usually been solved on the basis of a patently inappropriate application of the theory of voluntary responsibility. In this case, for instance, it may be said that X intentionally and voluntarily (with *mens rea*) served poisoned eggs to a human being in order to kill him. On the basis of this voluntary action—but by the use of a theory of status responsibility—X will be held responsible for the consequences of his illegal and immoral voluntary act, even though these consequences run counter to X's intentions and desires. Sometimes, of course, the law limits the criminal's responsibility to the foreseeable consequences of his act or to those consequences which a reasonable man in his position would have foreseen. (The law is not particularly troubled by the fact that a reasonable man either would not be in the criminal's position or, if he were, would no longer be reasonable.) But I feel sure that most lawyers, jurymen, and moralists would hold X morally and criminally responsible for the extension of his action in 17(*b*), despite the fact that X did not intend what happened, nor could he nor any reasonable man have foreseen these consequences. X would be blamed despite the fact that it would be morally objectionable to blame him on the basis of a reasonable and consistent application of the view that moral offenses must be voluntary.[46]

I firmly believe that X is responsible and blameworthy for what happened in 17. But the justification for holding X responsible must be on grounds of his status or ontological responsibility. X, like all persons, is dependent in action upon a matrix which may truncate, fulfill, or extend his action in such a way that his action is concretized in a way that may coincide with or be in disrelation to his plan or intention. But the action —as it comes to be whether in or out of accord with his intention and volition—is his action. There is no basis for crediting him with his virtuous actions in perfect relation to his intention unless his person and responsibility are enlarged to include elements of the ontological matrix over which he has at best limited control. Indeed, the ontological matrix is a part of his own person no matter how narrowly he defines it. His volition is never independent of the limiting conditions of his intelligence, knowledge, imagination, emotionality, and energy. Limit the person to what is under his voluntary control, and he disappears without trace along with his volition. If we acknowledge a sufficient number of ontological conditions, over which X has no voluntary control, to account for the existence of X and his capacity to act voluntarily, we have already acknowledged to a significant degree his moral responsibility for what he is no less than for what he does.

We extend this basic point only to a minor degree in recognizing the moral blameworthiness of X for *being* the man who voluntarily contributed to the situation in 13 that was transformed *without* his knowledge, intention, recklessness or negligence—that is *without* his voluntary participation—into the situation at 17(*b*). In 17(*b*) his voluntary act of killing his brother has been cut short prior to fulfillment. But in 17(*b*), X has the being or status of a murderer. He is the man who bears the volitional guilt of his brother's murder without the ontological guilt, and he is the man who bears the ontological guilt of his mother's death without the volitional guilt.[47]

I think the line of reasoning used by the court in *Suvada* vs. *White Motor Company* and related cases contains the elements of status responsibility infused with a trace of prior but not present voluntary responsibility and *mens rea* that I have in mind. X created the risk of poisoning someone other than his brother— however slight, non-negligent, and non-reckless that risk might be—in order to reap the benefits of his brother's death, just as the defendants in *Suvada* built, advertised, and sold for profit machinery which created an inadvertent risk. X would have gladly accepted moral praise for the philanthropic use of his brother's estate, had he inherited it; being rich would have given him the ontological basis for certain moral virtues which might be lacking without the estate. The defendants in *Suvada* may be well respected for the philanthropy which rests on the ontological basis of their risk-creating manufacturing and selling.

The consequences of X's action in 13 are outlined in 17. X is responsible for 17. 17 describes his action, not because of anything he did at 17 but because he *is* at 17 the man who *did* what 13 describes. We find the person of X in 17 only because in 17 he *is* the same man whose person was expressed in 13. Status responsibility rather than voluntary responsibility justifies our blaming him for 17. But his status responsibility in 17 derives from his mixed responsibility in 13.

Before considering a serious objection which Hart might raise to this line of reasoning, let us consider the subtle responsibility of X's brother in 17(*b*). He is the man who gave the poison to his mother. His position is like that of the mother who takes thalidomide as a sleeping potion and gives birth to a deformed child, or of a man who kills the child that suddenly runs in front of his car. In thinking of his role in his mother's death, X's brother is not tortured as the thalidomide mother must be with the thought that she bought her ease at the possible risk of harming her child. Nor is he troubled, as the unfortunate driver is, with the thought that by participating in a vehicular civilization and doing little or nothing to improve the safety features of our streets or cars, he has run the risk of killing a child. There are no antecedents to X's brother's action which contain any traces of voluntary personal involvement—none that might have contributed to or colored his action in 17(*b*). His volitional innocence, and his good fortune perhaps, leave him in status innocence as well. It is a precarious innocence, however, that could be compromised by his having thought 'What a disgusting old lady' just before he handed her the eggs intended for himself.

C

In offering my interpretation of the largely ontological basis rather than the voluntary basis of X's responsibility in 17, I have been troubled by an objection that Hart might reasonably raise. Suppose we were to ask X if he could establish the conditions which on Hart's view would excuse him from all moral blame in 17(*b*). X could certainly claim that 17(*b*) was completely unintentional in so far as he was concerned. But could X claim that he had taken every possible precaution? Would Hart say that there was one obvious precaution X had deliberately ignored—namely, the precaution of leaving the poison out of the eggs?

The objection looks formidable but its plausibility actually rests on the details of 17 and not on the issues of responsibility. Suppose, for instance, that X's brother had been Adolf Hitler and that 17(*b*) took place in 1943. We might all agree that X was an unlucky hero, like Colonel Stauffenberg, but could we fail to blame him, at least in some moderate degree, for his mother's death? When X takes upon himself the status of an assassin, as in 13 (assuming that the brother is Hitler), he establishes the ontological basis for his personal, moral involvement in what follows, even though it is not a voluntary or intentional or negligent consequent of his action in 13, and even though his action in 13—at the last moment it is to any degree voluntary— is morally good!

Perhaps the point can be seen with greater force if we consider a continuum in which the agent's immediate action is not morally or legally wrong or blameworthy. Following the conventional views of contemporary Americans and Europeans, let us suppose (*a*) that sexual intercourse between consenting adults of the opposite sexes is not morally offensive or blameworthy, and (*b*) that it is morally offensive and blameworthy to be the parent of a bastard child. If we take these moral suppositions for granted let us consider the following alternative situations in which— unless otherwise noted—Y is a healthy, young unmarried woman:

1. Y is raped and becomes pregnant.
2. Y is so seriously ill that advanced pregnancy would be fatal; she takes every contraceptive precaution, but becomes pregnant.
3. Y has no idea what causes babies, has intercourse without being aware of the possible consequences, and becomes pregnant.
4. Y with full knowledge of the cause of pregnancy takes all contraceptive precautions, has intercourse, and becomes pregnant.
5. Y with full knowledge is negligent in the use of contraceptives and becomes pregnant.
6. Y with full knowledge but in the ecstacy of love uses no contraceptive and becomes pregnant.
7. Y with full knowledge uses no contraceptive because of religious scruples and becomes pregnant.
8. Y with full knowledge, complete self-control, and no scruples against contraceptive methods, does not use them and becomes pregnant, not caring one way or the other.

This series is ordered on a continuum of undiminished or increasing volition as we move from involuntary to clearly voluntary action without extending the continuum to include any actions that involve the conscious intention to become pregnant.

Accordingly Y never intends to become pregnant, and in situations 1 through 4 she is neither reckless nor negligent. Can she not claim then that in so far as the first four situations are concerned, her action is morally blameless because unintentional and that she took every possible precaution?

Suppose Hart were to reply that she is confused about the action which requires justification. She intentionally had intercourse in all but the first situation. But if she did not provoke the rapist, she is not to blame for anything in 1. And since she took every precaution to avoid becoming an unwed mother, she has done nothing wrong in 2, 3, and 4, for there is nothing morally blameworthy, *ex hypothesi*, about intercourse. At this point Y will have to say: 'But Professor Hart, I am pregnant. As things stand now I shall become morally blameworthy for being the mother of a bastard child.'

How is Hart to cope with this situation on his theory of the voluntary character of moral offenses? The intercourse is not wrong, and on the voluntary theory of action and responsibility there is no act of becoming pregnant or of growing a baby. One does not become pregnant voluntarily or voluntarily develop a child in the womb. Hence, on this theory, there is no way that one can be morally blameworthy for becoming pregnant or producing a child except as a consequence of voluntary intercourse. But since we hold that intercourse is not wrong, we must then acknowledge the absence of any wrongdoing, if pregnancy follows when all precautions have been taken. One cannot admit that intercourse is morally right *per se* and then argue that in order to take all precautions against pregnancy one must refrain from intercourse. When one recognizes that intercourse is morally acceptable, he is committed on the voluntary theory to withholding moral blame for pregnancy if all precautions short of sexual abstinence have been taken. On Hart's theory we have the absurd consequence that no moral blame can attach to having a baby out of wedlock provided it was not planned and all precautions short of abstinence were observed.

204

The absurdity is even more glaring with regard to the moral blameworthiness of fathers who have children out of wedlock. Impregnating, like being impregnated, may be a highly personal act. But it can be so only on a view that takes seriously the notion of an organic mode of personal action and organic intentionality. Procreation cannot even be accomplished physically by the father; so far as he is concerned, it is an act by proxy, and the proxy is a germ cell completely detached from his body. If he has any effective intent in procreation beyond intercourse, it is by means of the organic intention of his proxy. To speak of his procreating voluntarily makes sense on an enlarged view of voluntarism which absorbs a considerable portion of Bergsonian organistic intention. But this Bergsonian view is not compatible with Hart's metaphysical asceticism. On his view the man who has intercourse without intending to have a child and who sees to it that proper contraception is used, must be free of all moral blame in fathering a bastard child.

Further evidence for the theory of responsibility I am urging is that it does not fall into this absurdity with regard either to Y or her male partner. On my view it is primarily the being of the father and the mother, not their voluntary act of intercourse, that provides the basis of their parental obligations and their blameworthiness for being parents out of wedlock. Like X in 13 and 17(b), the parent of a bastard child is responsible for being the person who set in motion factors that culminate in the development and birth of the child. There is a far greater degree of personal involvement in this situation, however, because of the organic expression of personality that is absent in the example of the poisoned eggs. The personal agents of the voluntary act of intercourse are present in procreation, not as chemical or mechanical forces, but as the living, organically intentional, fertilized egg with their personal genetic structure.

Both fathers and mothers must recognize and accept their responsibility for what they are and for what they become as a consequence of what they are, even when there is no bond created by intention or negligence that unifies what they are with what they become by reference to what they do.

Action broadly defined is that which binds past to present to future; it is the substance of personal duration. We destroy this bond or deny its existence by accepting a narrow voluntaristic definition of action and responsibility, and we lose the continuity of the moral self. Kant's old problem of providing some basis for moral continuity in a self which is unqualified in volition, except as it qualifies itself through the voluntary action in every moment, will be our problem unless we accept the necessary minimum of status elements in our concept of moral obligation and responsibility.

The adherents of voluntary responsibility cannot evade these difficulties by holding that Y, after becoming pregnant, is responsible only for doing whatever is necessary to avoid having the child out of wedlock; by holding, that is, that Y is excused from all moral blame if she makes every effort either to have an abortion or get married. It would be difficult to justify abortion in any but the first or second instances, and even the first case poses some problems if one recognizes intrinsic value in human life.[48] Nor is the alternative of marriage universally satisfactory: it may be impossible for Y to marry, or any possible marriage might be morally more objectionable than the offense of having the child out of wedlock. The father may have died; Y may conclude after careful deliberation that the father or any man whom she could marry would exert a morally corrupting influence on the child; or Y or the father may conclude that to marry would be so destructive of career to which an obligation is also owed that it is morally better to accept the blame for having the child out of wedlock without compounding moral offense by an immoral marriage. However we turn the problem around, there is no way of establishing the moral responsibility of Y or her partner for the unintended consequences of their non-negligent intercourse, without supplementing the voluntary theory of responsibility by the introduction of status elements; yet there is no way of denying their responsibility without abandoning our initial premise that it is morally wrong to have a bastard child.[49]

But suppose we take the argument one step further. Let us drop the supposition that it is wrong to be the parent of a bastard child—a dubious supposition despite its conventional support—and hold instead merely that it is wrong to be voluntarily, recklessly, or negligently the parent of a child for which reasonably adequate care and provision has not and cannot be made. With this revision, what shall we say of the expectant mother who has not been negligent or reckless in the use of contraceptives and who was reasonably confident of their adequacy. If moral offenses must be voluntary, she can be morally responsible only for future voluntary acts—for example, the neglect of her developing child. But our expectant mother may insist: 'Since I am not voluntarily (or morally) responsible for being pregnant, I refuse to alter my life because of this fact. And if I am blamed because I refuse to care for myself or my child, those who blame me must blame me in violation of Hart's principle that moral offenses are essentially voluntary. Those who blame me must blame me for refusing to meet a standard which, though appropriate to voluntarily expectant mothers, is not appropriate to an involuntarily pregnant woman like me. On the voluntary conception of moral responsibility, I am no more obligated to care for my child than to care for a wart on my nose.'

I assume that we agree that this pregnant woman has an obligation to care for the child she is carrying. But what reply can we make to her statement? If the thesis that moral offenses must be voluntary is sound, and if (having done her best to avoid pregnancy) she is not voluntarily pregnant, how can she be blamed morally for failing to meet a standard appropriate to her actual status (which she tried to avoid) but totally inappropriate to a non-pregnant status (which she voluntarily but ineffectually chose). In terms of the status and conditions she voluntarily chose, she is not required to care for her developing child. But because of a change in her status—through no choice or fault of her own—she is now obligated to care for it. She can claim, on the basis of the voluntary thesis, that since she did not choose the condition on which the new obligation is based, she does not choose to violate that obligation. Admittedly, she chooses to *act* in violation of the obligation to care for the child. But since she did not choose the condition on which the obligation depends, she does not choose to *violate* the *obligation*; hence, she violates no moral rule binding on her under the voluntary thesis.

Here we confront an essential feature of moral obligation which the voluntary thesis does not adequately account for— namely, moral obligation may obtain whether or not it is chosen. Moral obligation obtains according to the nature and

207

the situations of persons. While it is true that one's obligation may be changed by the degree to which he has voluntarily altered his nature or his situation (through education, recklessness, or contract, for examples), moral obligation normatively regulates his action whether his nature and condition are within his voluntary control or not.

In short, the context in which he is held to account for his voluntary actions (which are by no means purely but which, as I have shown, contain status elements) is an involuntary, necessary one. And if, in accordance with the voluntary thesis, one could excuse himself from an offending action by showing that he could not help it, it should be possible always to excuse oneself whenever he finds that he is of a nature or in a situation contrary to his voluntary control. For one can say: 'I cannot help being in this situation, and if I were not in this situation, my offending conduct—even if voluntary—would not offend.'

The fundamental basis of moral obligation is found in the nature and situation of the agent. These status elements are pre-conditions of the possibility of moral offense. And these pre-conditions are the consequences of prior voluntary actions only in some but by no means in all instances. The status elements, which define the agent's obligations, often arise without help from and despite the intentions of the agent; and they can never result merely from his voluntary choice. Hence, either we must abandon moral obligation as meaningless or we must abandon the voluntary thesis. Accepting the latter alternative, I hold (returning to our example) that the expectant mother is liable to judgment by an involuntarily imposed standard because she is in an involuntarily contracted condition of pregnancy. And I hold that her offense, if she violates the duty to care for herself and her baby, is not an *essentially* voluntary offense because it is defined by a standard which is involuntarily imposed on the basis of her involuntarily established condition.

The foundation of moral obligation (and therefore of moral offense) in status is fundamental. I may be privileged to argue that since I did not choose my skin color, I will alter it; or that since I do not voluntarily have a crooked nose, I will have it bobbed. But I cannot argue that since I did not choose my sex, I will have it changed;[50] or that since I did not choose to become a father, I will refuse to support my child.

208

JOHN R. SILBER

Either there are no moral obligations or moral obligations impose restrictions on my conduct because of my nature and my situation even if neither is a product of my voluntary choice.

D

Hart says, at the conclusion of the quotation we have been examining: 'in morals "I could not help it" is always an excuse, and moral obligation would be altogether different from what it is if the moral "ought" did not in this sense imply "can".' We have already seen the confusion and artificiality involved in trying to show that 'I could not help it', and I have argued that there are many situations in which 'I could not help it' is not an excuse. Consequently, I have argued that Hart's thesis of the voluntary character of moral offenses itself distorts the meaning of moral obligation. But the question remains whether, in rejecting the claim that 'I could not help it' as always an excuse, we have transformed moral obligation into something altogether different from what it is.

Two things must be said. First, even if we radically change the view of obligation defended by Hart, it does not follow that we shall alter the character of moral obligation—we may merely articulate its nature more precisely. This is not necessarily a logical issue about the meanings of words, despite the fact that Hart, and linguistic philosophers generally, so regard it. Second, in any case I have not altogether transformed the character of moral obligation for I have tried to present and defend the view that 'I could not help it'—though not always an excuse—would in many circumstances excuse one from the most severe degree of moral blame. One is excused from the degree of moral blame appropriate in cases of consciously voluntary wrongdoing provided one can show that what he did was not fully voluntary. On this point the traditional voluntaristic position on moral obligation is not altered.

But I have also argued that the application of ethical rules and principles in the absence of a theory of ethics containing: (1) a concept of responsibility which provides for a continuum of increasing and decreasing status and voluntary elements, and (2) a system of substantive values to overcome the abstractness of formalism, will never sustain a notion of moral obligation or

209

responsibility adequate to account for the subtle but extensive range of guilt and innocence, virtue and vice, praise and blame in personal action. Without these factors no notion of moral obligation or responsibility can be adequate to account for the fundamental moral virtue of accepting one's being or the fundamental moral offense of refusing to do so. And it is important to note that a part of the virtue of accepting one's being lies in the acceptance of one's partial blameworthiness for what one is.

Theories like Hart's or Kant's, which restrict responsibility to voluntary conscious acts and limit moral offenses to consciously intended or reckless acts in violation of principles or rules, can never give content or substance to moral action or to the moral person. Human choice is not something isolated from the choosing person. Rather it is a thoroughly organic mode of self-expression and self-discovery. There are gradations of choice and degrees of voluntariness; at every instant, however, even in those acts of purest, freest, most voluntary choice, choice depends upon the being of the person and the matrix of his action, both of which contribute to the moral quality of the person and to the moral quality of his action even though neither is to any great extent subject to his voluntary control. We shall never find a person in his action unless his doing is his being, or at least a part of it. We have neither understanding of man nor a basis for moral and legal judgment of him, unless we recognize the human person as a unity of being and doing. This cannot be done without partially reshaping the concepts of moral obligation and responsibility in order to free them of the limitations of simplistic voluntarism. We must reshape the concept of responsibility and the notions of intention, awareness, and choice so that they apply in varying degrees over the entire range of personal existence. When this reshaping is done, we shall have a better understanding of moral obligation and moral and criminal responsibility.

The voluntary conception of moral obligation and responsibility appropriately characterizes maximally personal actions. But an adequate conception must appropriately characterize *all* degrees and kinds of personal action from the most voluntary to the least voluntary. For this a theory of responsibility is required in which there is a fusion of status or ontological and voluntary elements, and in which moral obligation is understood

210

as applying not merely to the volition but to the being of each person.

Such a theory, when fully developed, will make necessary the reformulation of the judge's instructions to jurors in criminal cases. Jurors will no longer be asked to say 'Guilty' or 'Not guilty.' Rather jurors will be asked to find from among four to eight clearly formulated types of personal action, the type most nearly descriptive of the behavior of the accused. And on the basis of the jury's finding of the being-and-doing of the accused, a revised criminal code will prescribe punishment, treatment, or release. Throughout, however, a defendant in a criminal case will enjoy full protection of legal counsel and the procedural guarantees of the criminal law. For the recognition, in such a theory, of an attenuated element of *mens rea* and *actus reus* in the so-called status crimes of drug addiction and chronic alcoholism, etc., will preclude the withering away of the criminal law and the abandonment of those accused of such crimes to the less carefully controlled and perhaps less just procedures of medical boards.

When this theory is fully developed, it will account for the moral virtue of accepting one's being, and the moral blameworthiness of one's refusal to do so. It will make clear, moreover, that part of the virtue of accepting one's being lies in the acceptance of one's blameworthiness for what he is. We may even be able to understand the doctrine of original sin ('In Adam's fall we sinnèd all') as well as Oedipus' ultimate self-condemnation expressed in Sophocles' play by his act of blinding himself.

REFERENCES

[1] 'Status' here means the (legal) condition or state of being of a person, not, of course, his social status or prestige.

[2] J. Hall, *General Principles of Criminal Law* (Second Edition 1960), pp. 70, 179, 230, 251.

[3] H. L. Packer, '*Mens Rea* and the Supreme Court,' *The Supreme Court Review* (1962), pp. 107–9, cf. American Law Institute, Model Penal Code, Tentative Draft, No. 4, Comments, § 2.05 (1955).

[4] H. M. Hart, 'The Aims of the Criminal Law,' 23 *Law and Contemporary Problems*, 401 (1958).

[5] H. L. A. Hart, *The Concept of Law*, p. 173 (1961).

[6] *Ibid.*, pp. 173–4 (1961).

[7] 370 U.S. 660, rehearing denied, 371 U.S. 905 (1962).

[8] Section 11721 of the California Health and Safety Code.
[9] 370 U.S. at 662.
[10] *Ibid.* at 666.
[11] *Ibid.*
[12] *Ibid.* at 667.
[13] *Ibid.* at 669.
[14] *Ibid.* at 674, 676, 677 (footnotes omitted) (emphasis in original).

Mr. Justice Harlan, concurring, did not agree that narcotics addiction is an illness nor, consequently, that to subject narcotic addicts to criminal sanctions would amount to cruel and unusual punishment. But, like Justices Stewart and Douglas, he denied the right of the State to convict a person for his addiction to narcotics rather than for their use. Since, according to Justice Harlan, addiction is not more than 'a compelling propensity to use narcotics' (370 U.S. at 679) he reasoned that the California court had in effect authorized 'criminal punishment for a bare desire to commit a criminal act'. (*Ibid.*) And he refused to permit the substitution of a wish following from status for an action, an *actus reus*, in the definition of a crime.

[15] But the reluctance of our courts to use the concept of status responsibility in criminal cases has not been universal. Many states using the common law have applied criminal sanctions to the offense of vagrancy. Prior to *Robinson* vs. *California*, a vagrant could be anyone, from a healthy beggar,[a] to a prostitute,[b] to a narcotics user,[c] it is clear that such laws punish a state of being or a condition and abandon the requirement of conduct, *actus reus*, in the definition of the criminal offense. Confinement for vagrancy is a punishment, in the words of Justice Holmes, 'for "being a certain kind of person not [for] doing a certain overt act. . . . It follows . . . that . . . the conduct *proved is not* the offense but only a ground of inference.'[d] But not only does the decision in *Robinson* vs. *California* stand in opposition to crimes of status; this opposition is likewise found in the *Model Penal Code* and the Uniform Narcotic Drug Act. ([a]Ala. Code Title 14 § 437 (1958); [b] Tex. Pen. Code art. 607 (1952); [c] N.J. Rev. Stat. § 2A: 170–8 (1953) (disorderly person); see *Dubin & Robinson. The Vagrancy Concept Reconsidered*, 37 N.Y.U.L. Rev. 102, 109–13 (1962) (exhaustive list of categories); [d] *Commonwealth* vs. *O'Brien*, 179 Mass. 533, 534, 61 N.E. 213, 214 (1901) (emphasis supplied).

[16] In *United States* vs. *Dotterweich*, 320 U.S. 277 (1943), the Supreme Court recognized the validity of a law which, in the words of the Court, 'dispenses with the conventional requirement for criminal conduct—awareness of some wrongdoing' (320 U.S. at 281). Here the court upheld the criminal conviction of Dotterweich, the president of a company that had made two interstate shipments of drugs that were either mislabelled or adulterated, although it had not been shown that Dotterweich was either personally aware of the mistaken shipments or negligent in his administration of the company.

And in *United States* vs. *Balint*, 258 U.S. 250 (1922), and in *United States* vs. *Behrman*, 258 U.S. 280 (1922), the Supreme Court so construed the Harrison Narcotics Act that knowledge that one was selling narcotics was not made an element in the offense of selling them. And the Court raised no objection to the provision of the five years' imprisonment as the maximum penalty under the statute for an offense which could be proved against a person who did not knowingly engage in the activity proscribed by the statute. (It should be noted that the issue of imprisonment was not raised on appeal by the parties in *Balint*.)

[17] See e.g. H. L. A. Hart, *op. cit.*, pp. 168–9, 173–5; H. M. Hart, *op. cit.*, 401, 422–5; J. Hall, *op. cit.*, pp. 342–51, 375; H. L. Packer, *op. cit.*, pp. 109–10, 147–8; American Law Institute, *Model Penal Code*, Tentative Draft No. 4, § 2.05 and Comments, p. 140; F. B. Sayre, 'Public Welfare Offenses,' 33 *Columbia Law Review*, 55 (1933); R. A. Wasserstron, 'Strict Liability in the Criminal Law,' 12 *Stanford Law Review* (1960), p. 731; G. Williams, *Criminal Law* (1953), §§ 70–76, 81.

[18] This is not to deny that some acts which give rise to tort litigation do involve the moral blameworthiness of the agent.

[19] *Suvada* vs. *White Motor Company*, 32 Ill. 2d 612, 210 N.E. 2d 182 (1965).

[20] *Ibid.* at 618–19, 210 N.E. 2d at 186. I inserted [*a*], [*b*], and [*c*]. I wish here to thank Professor Wayne Thode of the University of Texas School of Law for informing me of the extensive use of strict liability in tort law, and for his general criticisms of this paper.

[21] Justice Harlan, as we noted, *supra* note 14 avoided this pitfall in his concurring opinion.

[22] This appears to be the direction taken by the Circuit Court in *Driver* vs. *Hinnant*, 356 F. 2d 761 (4th Cir. 1966). The Court held that the Constitutional provision against cruel and unusual punishment precluded North Carolina's punishing a chronic alcoholic for public drunkenness. In language reminiscent of *Robinson* vs. *California*, the Court said, 'The upshot of our decision is that the State cannot stamp an unpretending chronic alcoholic as a criminal if his drunken public display is involuntary as the result of disease' (356 F. 2d 765). 'The alcoholic's presence is not his act, for he did not will it. It may be likened to the movements of an imbecile or a person in a delirium of a fever' (356 F. 2d 764). See also *Easter* vs. *District of Columbia*, 361 F. 2d 50 (D.C. Cir. 1966), in which the Circuit Court recognized chronic alcoholism as a defense to the charge of public intoxication.

Future developments may be presaged in the dissent of Mr. Justice Fortas to the Supreme Court's denial of *certiorari* in *Budd* vs. *California*, (*Budd* vs. *California* cert. denied, 385 U.S. 909, 912–13 (1966)). 'Our morality does not permit us to punish for illness. We do not impose punishment for involuntary conduct, whether the lack of volition results from "insanity", addiction to narcotics, or from other illnesses. The use of the crude and formidable weapon of criminal punishment on the alcoholic is neither seemly nor sensible, neither purposeful nor civilized.'

[23] This is the same passage quoted at note 6 *supra*. Hart's original phrases are inserted in brackets following my italicized alterations.

[24] The relevance of this discussion should be apparent in due course. If responsibility is limited to pejorative contexts, it will be difficult if not

P   213

impossible to present in ordinary English a continuum theory of responsibility in which voluntary and status elements are combined. The increased difficulty would be, moreover, the gratuitous consequence of linguistic dogmatism.

[25] Here, of course, 'intentional' has a common sense and not a phenomenological, technical meaning.

[26] I shall consider later the question of whether the acceptance of this view alters the relation of implication between 'ought' and 'can' in such a way that, as Hart alleges, moral obligation would be transformed into something altogether different from what it is.

[27] S. Hampshire, *Thought and Action*, London: Chatto & Windus, 1959, p. 94. Consider also Hampshire's statement: 'A more decisive difference between consciousness and unconsciousness lies between the necessity of intended action in one case and the mere natural movement without intention in the other' (*ibid.*). When done by linguistic fiat, as in this instance, philosophy becomes as easy as it is irrelevant.

[28] Mircea Eliade lays great stress on the National Socialists' selection of the Nordic myths as an expression of national purpose: the goal is *ragnarök*—total destruction of gods, heroes, and men! See *Myths, Dreams and Mysteries* (P. Mairet, trl.), London: Harvill Press, 1960, pp. 26-7.

[29] When Chamberlain said that he brought back from Munich 'Peace for our time,' he let the cat out of the bag. It is not hard to find the intent behind that phrase or behind the equally famous '*Après nous, le déluge.*'

[30] *State* vs. *Blomquist*, No. 33391 (D. Tex., May 17, 1962).

[31] Survival of the institution of marriage would require the negligence of most of the adult population if such a consideration were made a rule of law.

[32] We will discuss later the problem of accounting for negligent conduct on a theory which recognizes only a conscious level of intention and awareness. See Marshall, *Relation of the Unconscious to Intention*, 52 V.A.L. Rev. 1256 (1966).

[33] Hart would not necessarily excuse him from legal responsibility. Hart would not accept without serious qualifications Jerome Hall's position that 'Penal law implies moral culpability' (Hall, *op. cit.*, p. 347).

[34] Here I am making a theoretical point about law and morals. Practically speaking, it will be difficult, perhaps impossible, to prove 'beyond reasonable doubt' what a man has thought or dreamed, etc. Some move might none the less be made in this direction. It must be emphasized, moreover, that the introduction of such considerations might be made for the purpose of exonerating a person or for mitigating his guilt. It would be a serious mistake to suppose that the view I am developing tends more to incriminate than to exonerate mankind. My argument does not increase one's moral and legal responsibility; rather, it attempts to redefine and clarify the nature and scope of responsibility.

[35] I have indicated at several points in this paper that Hart's statement represents, or is representative of, the generally accepted view of the voluntary character of moral offenses. I selected Hart's statement for examination because it epitomizes the view that has been dominant among ethical

writers from Aristotle to Kant as well as because of its clarity and brevity.

[36] A.L.I., *Model Penal Code*, § 2.02, Comment (Tent. Draft No. 4, 1955).

[37] *Ibid.*, p. 126.

[38] In my consideration of the case in which the Texas man killed his wife (*State* vs. *Blomquist*, No. 33391 (D. Tex., May 17, 1962), 1962), I dealt superficially with the question of his possible 'recklessness or negligence' because I had not yet introduced the technical distinction between negligence and recklessness. In retrospect it should be clear that the theory as represented by Hart is reduced to an absurdity when applied to this case because the man's conduct was morally excusable since it was, in terms of the theory, neither intentional, negligent, nor reckless.

[39] Professor Herbert Packer, *op. cit. supra* note 3, has urged the consideration of negligence as a conceptual halfway house between strict liability and *mens rea*. If we accept as factual a side of mental life of which we are not directly conscious but which, according to many psychologists and psychoanalysts, accounts for slips of the tongue, deliberate forgetting, and other failures that can be grouped under the general heading of negligent behaviour, a substantial element of the *mens rea* requirement could be reintroduced into criminal and tort law under rules concerning negligence at points where at present rules of strict liability are used or where negligence is treated as if it were devoid entirely of *mens rea* and hence where rules of negligence are applied exactly as rules of strict liability.

[40] A continuum theory of human conduct and responsibility may have been behind and is certainly required by the A.L.I. proposal, in § 2 of the *Model Penal Code*, of four modes of culpability, which in descending order are: purpose, knowledge, recklessness, and negligence.

[41] My discussion of these thirteen situations will of necessity be very brief and sketchy; it should serve, nevertheless, to carry the reader on his own through many of the considerations which I find relevant.

[42] S. Freud, *Some Additional Notes Upon Dream-Interpretation as a Whole: (B) Moral Responsibility for the Control of Dreams, Collected Papers*, Vol. V, p. 154 (Strachey ed. 1959).

[43] See my 'Human Action and the Language of Volitions,' *Proceedings of the Aristotelian Society*, Vol. XLIV (1963–64), pp. 199–220.

[44] We raise this question while recognizing that to *will* something, unlike merely to wish it, involves a determined effort on the part of the agent; within the limits of his capacities, the agent does everything he thinks is required for the fulfillment of his intention. If one *wills* something to happen, its failure to come to pass cannot be imputed to the volition of the agent for if he wills it he does what he can to bring it about and does nothing to prevent it. As the article just cited, *supra* note 43, will show, I do not accept any para-mechanical theory of the relation of mind to body or any of the other horrible things that Ryle alleges are accepted by those who use the language of volitions.

[45] I use the word 'fortunately' in order to assert again the shocking fact that luck, accident, or fortune play important and partially determining roles in the shaping no less than in the assessing of moral responsibility.

[46] In law one might apply the fiction of comparable intent; such a move

215

is, of course, patently inadequate. Legal fictions are simply *ad hoc* corrections to defective legal theories.

[47] By recognizing X's mixed responsibility, we can assess the full quality of his mother's death, despite the fact that we lack the precise noun, adjective, or adverb to articulate its quality. Her death was not quite a murder, but neither was it mere manslaughter; certainly it was not accidental homicide.

[48] I suppose it can be argued that the neglect and abuse to which a bastard child is often subjected can be avoided by killing the child. The legitimacy of the argument might be more apparent, however, were it urged by the illegitimates. My experience is that natural-born bastards are just about as intent on living as self-made ones.

[49] The degree of Y's status responsibility in having a child out of wedlock will vary in the situations from 5 to 8 according to the degree of voluntariness and the extent of negligence or recklessness in the use of contraceptive devices. These further considerations are left to the reader.

[50] *Anonymous* vs. *Weiner*, 50 Misc. 2d 380, 270 N.Y.S. 2d 319 (Sup. Ct. 1966) (denial of application to change sex designation on birth certificate to correspond with results of medical operation which changed petitioner's sex). See H. Benjamin, *The Transsexual Phenomenon* (1966) for a medical discussion and defense of the right to alter one's sex. But Dr. Benjamin supports surgical transformation of sex in only very unusual situations where there is a serious psychological disturbance in gender role and gender orientation. See also Benjamin, *Clinical Aspects of Transsexualism in the Male and Female*, 18 *Am. J. of Psychotherapy*, 458 (1964); and Benjamin, *Nature and Management of Transsexualism: with a Report on 31 Operated Cases*, 72 *Western J. Surgery, Obstetrics & Gynecology*, 105 (1964).

PART II
PHILOSOPHICAL PERSPECTIVES

8

PHILOSOPHY AND CIVILIZATION

Newton P. Stallknecht

This paper is written in defense of a few platitudes, the basic platitudes of our civilization. These platitudes are harder to accept and to defend than many of their sincere supporters have recognized. They are thus not really platitudes at all, but—shall we say?—'plongitudes' disguised by their own prestige. As this becomes obvious they may regain their rightful place at the heart of philosophical discussion.

The sense of value that supports the ideology of our Western world has long centered upon the idea of the autonomous and responsible individual. This notion, as a regulative idea, never completely realized, characterizes our civilization, and its reflection in religion, social policy, and the arts is often considered an index of our cultural maturity. From this point of view, such an institution as school, church, or state, indeed any corporate entity, can ultimately justify itself only in so far as it enriches the lives of its members, and of those who come in one way or another under its influence, by granting and encouraging freedom of self-development.

Furthermore, we have come to recognize that the quality of these organizations depends upon the contributions of individual human beings whose thinking and overt action revitalize and redirect the cultural movements upon which their effectiveness depends. It is from the insight, decisions, and commitments of its individual members—both of those who attain leadership and those who support the leadership of others with sympathy and understanding—that the health of our community is maintained. As William Wordsworth once put it, the true wealth of nations lies in the character of its citizens.

Before such a way of life can approach its full realization,

219

the notion of the individual both as an effective agent and as an end in himself must be clearly envisaged. Such vision is a considerable achievement. In many primitive cultures, even in that reflected in the Homeric poems, this notion is not fully developed. In Homer the gods, symbolizing or accepted as identical with forces of nature or deeply rooted tendencies of human nature, are often said to overwhelm the individual in moments of great crisis, when his decision can hardly be described as his own. In more primitive cultures, social attitudes and patterns of approved behavior actually leave the individual a very minimum of freedom. The fear of violating firmly established folkways and prohibitions stands as the primary source of motivation, and the individual, rather than being recognized as an end in himself and as a source of evaluation and decision, is held, so to speak, in solution within the life of the community. In contrast, the discovery—we might almost say the creation—of the individual, recognized both as an agent and as an ultimate value, stands as a revolution of enormous significance. Such a revolution is not a matter of a decade, a generation or even a century. It moves slowly between remote extremes, between, say, the unchallenged expulsion of a native who has, perhaps unwittingly, violated a tribal taboo and, on the other hand, the meeting of minds achieved by a group of voluntary fellow workers committed to a common objective, who have learned to share and mutually profit by one another's experience and insight.

We cannot here trace the notion of the individual, as an idea and as an ideal, to its historical origin. Let us point, however, to its early development in classical civilization, in the imagination and the thinking of the Greeks, where human consciousness came gradually to recognize itself as a source of self-determination. This notion is present, at least by anticipation, in ancient tragedy, and it takes on an intellectual form during the lively disputes between Socrates and the Sophists. It is often central in Plato's thought, as in the beautiful myth of Er, and again in Aristotle's sober consideration of moral decision. It is dominant in the thinking of the Stoics. Against a different background, it becomes increasingly clear in the development of Hebrew-Christian thought, gradually overcoming the heavy resistance offered by the institutionalism of religious establishments and the vested interests of powerful minorities. Springing from the heart of Christian belief, supplemented by Greek wisdom, this notion has in modern times inspired the internal organization of the Christian community, reforming ecclesiastical policy and re-orienting political theory and practice. It has also appeared in the arts and in education as the ideal of creation has gradually superseded that of imitation. One can hardly overemphasize the far-flung influence of the Socratic *Know thyself*, of the Christian 'As a man thinketh in his heart so is he,' and of the equally important 'The Sabbath was made for man, not man for the Sabbath.' In sympathy with these insights, we all respect the ideal of a kingdom of ends that furthers the realization of human individuals in a life of self-expression carried on in a spirit of cooperation and mutual responsibility. We are all ready to accept the ideal of such responsible freedom shared in community. For Western thought this freedom is often accepted as the very quality of man. In this sense we are all humanists. Even the Marxians have recognized such an ideal as justifying the rigorous restrictions of a prolonged period of social transi-tion. Indeed, so widespread and so authoritative has this ideal become that its statement is now little more than an obvious platitude and there is always the danger that we will take it too readily for granted as a background that we need not examine very carefully. As a result, this noblest of ideals is often caricatured by hasty, if sincere, thinking. Thus the notion of responsible individualism has at times been distorted to justify an economy of unrestrained competition, including the unrestricted freedom of an employer in his relations with his workers.

There are, however, difficulties of another sort that arise when we concentrate our attention upon the idea of the free indi-vidual. It has not been easy for modern philosophy to come to terms with this way of thinking. As a result academic philo-sophy and what we might call the traditional common sense of our civilization have often been at odds. It is an irony of history that as the ideal of the responsible individual gathers prestige in the political and religious thought of modern times, the concept of nature that emerges with the beginnings of modern science renders this ideal increasingly difficult to formulate in philo-sophical terms. Thus, when considered conscientiously by the

221

theorist, the freedom of the individual becomes difficult to define and especially difficult to reconcile with other concepts that the philosopher is often reluctant to dismiss.

As a physical organism, even an organism of extraordinary development, man remains a part of nature. Nature is interpreted as an interplay of events that manifest, at least on the level of molar masses and living things, a system of predictable patterns, including cyclical routines, so that the structure of the present appears as an extrapolation of the structure of the past. This would seem to leave no room for conscious initiative, for self-motivated control or telic orientation. Modern science, especially in the form taken by early developments in physics and astronomy, that is, by those very developments which in the eighteenth and nineteenth centuries appealed most vividly to the popular imagination, has been often inclined to identify the intelligible with the predictable and to insist that any genuine aspect of nature will in time be so understood. Human conduct is often included under this assumption. Many philosophers have accepted this bold postulate and have offered the layman an image of human nature that he has been reluctant to accept, preferring rather to view the philosophical enterprise itself as suspect.

The layman often stands closer to the spirit of Western civilization than the modern philosopher, and he is not ready to accept a theory of selfhood that renders our conscious motivation primarily an outcome of previous events, and therefore, ideally speaking, essentially predictable. For the layman, as for certain Greek philosophers, the psyche initiates motion, the human self stands as a center of choice that directs commitment. Sentences such as 'I changed my mind' or 'I kept my temper' must be interpreted in the light of these assumptions. This attitude of the layman is not a matter of sheer sentimentality. His opposition to a determinist interpretation of human behavior deserves more attention than many a philosopher is likely to bestow upon it.

Determinism can have grave consequences for human relations, consequences that may undermine the traditional respect for the individual that the layman finds so important. After all, if we consider the consciousness of our fellow man as the outcome of external conditioning, we may well come to follow the

222

strategy and techniques so often employed by the advertising expert rather than those of the Socratic teacher. We may well be tempted to control behavior rather than to share experience and exchange points of view. Once we have finally accepted a philosophy of determinism, we will not seek to awaken our fellows to problematic situations, inviting them to answer our questions in their own way and to offer suggestions that occur to them as they survey the situation from their own point of view. As teachers, we will be more interested in the formation of habits, perhaps primarily habits of speech, than in awakening a sense of responsibility. We will not be eager to share problems but to press toward the propagation of readymade attitudes. In such a situation, the teacher becomes a trainer, and the politician a salesman.

On the other hand, just in so far as we scorn, as the good teacher so often does, to employ such methods, we will find ourselves acting on assumptions that commit us to a very different concept of human nature. It is interesting to notice that the spirit of common courtesy, which reflects a genuine respect for the autonomy and responsibility of our fellows, renders distasteful any effort to press them toward practical commitments and overt statements of opinion that they have not made their own. In this respect we are all laymen and resent the slightest hint that our thinking is to be conditioned by expert technicians, whether they be salesmen, political or religious propagandists, or aggressive educators. The essence of courtesy lies in a willingness to respect the experience and insights of other people. This respect for others goes hand in hand with our own self-respect, for if we treat the thinking of others as the outcome of external conditioning, we can hardly make an exception of ourselves. Our own thought has been conditioned, too, even if we know not how. It is perhaps only in a community where the ideal of individual responsibility and autonomy is kept constantly in mind, both in theory and practice, that genuine thinking and sincere expression can flourish and be taken seriously.

The delight that we take in the arts and the respect that we pay the achievements of science include at least a tacit recognition of the freedom or responsible autonomy that is manifest in their production. This freedom has often to resist early conditioning rather than to profit by it. The poet or any serious writer must free himself from the use of clichés. Thus, he must resist some of the most powerful associations established by listening to the speech around him. He must even take care not to imitate his favorite authors. And the scientific thinker must be capable of surveying his data from many points of view, resisting at least temporarily the pressure of dominant opinion, as he explores unfashionable possibilities of interpretation. Successful work in art and in science requires an integration of detail that is possible only to a mind capable of resisting distraction of all sorts, a mind capable of reconditioning itself as it enters new situations and accepts new objectives.

But freedom is by no means negative in spirit. The ability to resist fashionable patterns of thought and commonplace modes of expression must be supplemented by the responsibility that underwrites or answers for (*respondeo*) those patterns and modes actually accepted and embodied in action. We all accept the proposition that such responsibility is possible. To deny it would be to consider the words *true* and *false* as uttered by human beings to be virtually meaningless terms.

The above seems, I think, to indicate that the layman, the artist, and the scientist have, all three, reason to respect the ideal of the autonomous and responsible individual, in whose conscious activity freedom of action, of expression, and of thought may be realized. From this point of view the nature of consciousness and its relation to its world assumes great importance, and it is to such inquiry that the philosopher should turn his attention. In studying consciousness, we find ourselves reminded of Plato's attitude in the famous Seventh Letter. We can often do little more than invite our readers to look for themselves. Verification will be in the self-examination of each one of us. At times a few steps of deductive argument will be possible. But in general we are appealing to the experience of the reader and urging him to report on certain features of his own conscious life.

Our consciousness presents itself as activity—not a Cartesian *res cogitans* or *chose qui pense*, but an activity that shapes itself so that it is difficult to tell the 'dancer from the dance.' And yet the dance offers a better metaphor than William James's famous 'stream of consciousness.' This is because in consciousness there seems always to be not only some sense of continuity with the past but some sense of direction or expectation and some sense of selective choice. We do not, to be sure, perceive an agent imposing direction upon our thinking. Rather some sense of continuity, some degree of choice and intention seem to characterize each moment of consciousness and to distinguish it from sheer feeling. These features vary with the intensity of consciousness. Together they afford a context within which the first person pronoun acquires meaning.

The basic activity of consciousness has to do with attention. We pay attention and we shift attention. Even a voluntary decision centers, as William James used to teach, upon a concentration of attention that initiates overt action. It is important to notice that attention is always both objectively and subjectively oriented. In the first place, consciousness, as both Kant and Husserl would insist, is always consciousness *of*. There is always reference to something, not to be considered as identical with consciousness itself. But at the same time we must recognize that the structure of consciousness justifies our use of the first person pronoun. Whatever object we apprehend, there is always some felt continuity of our present activity with our past and some sense of some further activity to follow. Acts of attention appear as *our* acts just in so far as they are not isolated from one another. Indeed, the subjective orientation of consciousness lies precisely in overcoming or avoiding this isolation. '*I am*' seems usually to be an elliptical expression. It seems to mean something as follows: *Here I am* doing thus and so, involved in such and such an undertaking begun in the past and opening upon a future still indeterminate in detail. In moments of reflection this time span referred to may be a very long one, half a lifetime or more, and still open to the future. This subjective continuity is maintained only so far as we recognize some relevance of present things and events to those things and events that preceded them and with which we were previously concerned. The transtemporal relevance is apprehended through our ability to interpret our situation as in some way continuing features of the past. This primary interpretation is basic to perception itself, which entertains objects and continued events by recognizing their appearance and reappearance over considerable lapses of time. When we remember that these appearances do not necessarily resemble each other, the importance of this primary interpretation becomes clear.

Consciousness is, however, not merely a matter of seeing our present as in one way or another a continuation of our past. We are always concerned in some degree with the future. There is something ever more about to be with the shaping of which we find ourselves involved; for consciousness is not purely cognitive: it has always a conative aspect. *I was, I will be* and *I want to be* or *I intend to be* qualify the *I am* which is no mere punctual presence. We can be misled by the neatness of the grammarian who finds *I am* and *it is* belonging to the same tense. In this respect *am*, along with the experience that justifies its use, stands in sharp contrast with *is*. *Am* straddles all three major tenses. From this point of view *am* and *is* should not be considered forms of the same verb. *Am*, so to speak, always projects itself into an unfinished and partially indeterminate future and carries its past along with it. Consciousness overlaps past and future. This is especially obvious in any activity that requires the employment of a language or a system of symbols. The construction of an intelligible sentence, to say nothing of a sustained argument, calls for many decisions and adjustments over which presides our conscious intention to communicate or to persuade. Here we must keep ahead of ourselves, sensing the direction or intent of statements still in the making. This unrealized intention constitutes our hold upon the future. This future may actually come to be in the decision that we are making, as we move from a scheme of alternative possibilities toward a texture of fully detailed and concrete events. While the alternatives are still open, consciousness is engaged in shaping an outcome. Thus a conscious individual lives largely in his own future, what we might call his future concrescent.

It is most important to recognize that this future concrescent is not a mere idea of ours. It has a reality of its own. It is not to be identified with our thinking. It is something about which we think and with which we work. Our thinking concerning this future may be true or false—false when in confusion we commit ourselves to an impossibility, as when we become circle-squarers or dream of constructing a perpetual motion machine —again false when we envisage a genuine possibility but fail to notice that we have no facilities at *our* disposal to assure *our* realization thereof. When, however, our thinking is free from any such confusion the possibility envisaged is a genuine component of our open future, even if at last we renounce it or fail to realize it. Of course, such an entity is not to be confused with a concrete object, but it remains a genuine object of responsible thinking, in this one respect similar to a Platonic form. Consciousness operates between what is and what may be, somewhat as the soul in Plato's philosophy apprehends both things and the forms. Consciousness is neither an actual concretion nor an unrealized possibility but a productive interplay of the two. Thus consciousness is an ontological amphibian. It qualifies and in a sense criticizes the concrete by setting it in a context of nonexistent possibility. This is supported by that triumph of conscious activity, the negative judgment, through which consciousness may invite the realization of what would otherwise never take place.

So conceived, consciousness, and with consciousness the individual selfhood that characterizes its higher development, cannot be identified with or located in the spatial layout of an organism as it exists at any one moment of observation. In this Leibnitz was right. If the human brain and nervous system were expanded in size to equal that of a mill or factory and we were allowed to walk within it, we would observe nothing but moving parts, of one kind or another, in their intricate relationship. We would not come face to face with feelings, perceptions, and ideas. This is consistent with the notion that consciousness is not limited to the concrete.

It does not exist as an actual event with a concrete tissue of other actual events. A pronounced syllable or word may so exist but not the consciousness that expresses itself through that word with reference to the meaning that is taking shape as the word is chosen. Here we are tempted to borrow a phrase from the idealists of an earlier generation. Consciousness, while in contact with the spoken word, *transcends* it. The meaning of *transcends* may be discerned through an examination of the conscious situation as a whole. A word, considered not as the beating of the air with the tongue but as enjoyed by consciousness, participates in a meaning, still incompletely formulated, with whose final formulation consciousness is concerned. The pronounced word may perhaps be likened to a point at which

227

activity of consciousness contributes to the concrete but does not identify itself with it.

Fully alerted and active consciousness, indeed any activity over which such consciousness presides, is not, like certain phenomena in the physical order, a system or series of events whose most prominent features may be considered as fully predictable. Consider a moment of choice—perhaps we should say of creative choice—when a poet adds a final phrase to a stanza, or corrects the ending of a line already in context, a moment in the growth of an unfinished poem. To some slight degree the new phrase has been predictable—by the poet himself or by a friend who has heard the lines already composed. This prediction may be based on the fact that the metrical scheme of the phrase and the rhyme tone of its last syllable must fit into patterns already established, and syntactically the phrase must conform with a sentence structure already in being. In other words, the new phrase must be reconciled with certain features of past composition, and hence its character is in some degree predictable. Yet the phrase itself considered as a fully concrete achievement, the phrase with the final wording that supports its imagery and metaphorical significance, has not been predictable. To predict to this extent would have been to complete the phrase and to have enjoyed its aesthetic relation to the poem as it took shape. Such prediction would coincide with creation. The predicting observer would have taken over the writing of the poem for himself. The first author, while perhaps admiring his observer's contribution, would in all probability not accept it as his own—unless he stood in deepest sympathy with the observer and was willing to accept him as a collaborator.

What is so obviously true of artistic creation should be true also of any act of fully alert consciousness that carries constructive choice beyond the consideration of familiar generalities. Prediction cannot reach the full detail with which the conscious individual realizes a long-term objective of recognized importance. To be sure, certain very general anticipations of human decision often take place. Thus we may be very sure that certain people will not choose to travel by air, and we may be well aware of the circumstances that have conditioned this decision. But just how these people will choose to travel as they plan and enjoy their vacations soon surpasses the scope of our foresight. What transport they will prefer from time to time, what will be their routes, their stopping-places, when they will be willing to exceed their budgets, etc., etc.—all this is not a matter of close prediction.

After all, to understand the activity of a conscious individual does not depend upon an effort to reduce its pattern to a trajectory. In so far as such understanding is possible, it calls for the methods of the historian and the biographer, even of the literary critic, and it is to their methods rather than to those of the physicist or the astronomer that we should turn in trying to comprehend the action and utterance of a human individual. What we seek is an intellectual sympathy through which we reproduce the orientation, including the evaluation of circumstances, available alternatives and objectives within which the individual has lived, has taken action, and has tried to justify it. We need, more than anything else, some glimpse of the individual's image of himself or of his work and his sense of what he would achieve. Such understanding can complete itself only in retrospect. This is very clear in the work of the great artist whose new departures often for a time baffle his most friendly critics.

It is, I submit, one of the chief tasks of the philosopher to explore the presuppositions of such inquiry. With such a philosophy human consciousness must occupy the central, indeed a privileged, position. But there is a danger that we will so emphasize its importance that we will isolate it by interpreting it as a miracle that terminates all further inquiry. Here we may profit by the wisdom of that master of speculative philosophy, the late A. N. Whitehead, whose contribution is today all too likely to be overlooked. Consider the following passage from Whitehead's *Modes of Thought*:

> Human nature has been described in terms of its vivid accidents, and not of its existential essence. The description of its essence must apply to the unborn child, to the baby in its cradle, to the state of sleep, and to that vast background of feeling hardly touched by consciousness. Clear, conscious discrimination [including the entertainment of alternatives] is an accident of human existence. It makes us human. But it does not make us exist. It is of the essence of our humanity. But it is an accident of our existence.[1]

Here, I think, a slight qualification is in order. Clear conscious discrimination is an accident of our *existence*, an accident, of course, in the Aristotelian sense. But it may become, even so, the heart of *our* individual existence—of the existence of a human being who achieves a responsible autonomy. This autonomy or mature individuality is not forced upon us. It is made possible but not guaranteed by our intricate biological and physical sub-structure and by a welter of feeling and unenlightened emotion. Without this sub-structure there could be no conscious *life*—nor would consciousness be in contact with its world. Here, perhaps, another quotation from Whitehead's book is in order:

I find myself as essentially a unity of emotions, enjoyments, hopes, fears, regrets, valuations of alternatives, decisions—all of them subjective reactions to the environment as active in my nature. My unity—which is Descartes' 'I am'—is my process of shaping this welter of material into a consistent pattern of feelings. The individual enjoyment is what I am in my role of a natural activity, as I shape the activities of the environment into a new creation, which is myself at this moment; and yet, as being myself, it is a continuation of the antecedent world. If we stress the role of the environment, this process is causation. If we stress the role of my immediate pattern of active enjoyment, this process is self-creation. If we stress the role of the conceptual anticipation of the future whose existence is a necessity in the nature of the present, this process is the teleological aim at some ideal in the future. This aim, however, is not really beyond the present process. For the aim at the future is an enjoyment in the present. It thus effectively conditions the immediate self-creation of the new creature.[2]

For Whitehead, we must remember, the 'present process' is not a static instant but an invasion of the future. Whitehead has boldly spoken of this process as a self-creation emerging from our 'subjective reaction to the environment.' He has even dared to use Spinoza's phrase *causa sui*. A finite *causa sui* is an even more difficult concept than that of Spinoza's *deus sive natura*. The interplay of mutual adjustment constantly taking place in what is sometimes called the concert of the living organism gives us some inkling of what is meant. Enlightened consciousness perfecting such an organism presents a more spectacular example. Here we have not so much an adaptation

230

to environmental circumstance as a control, within limits, of the environment itself—an imposing of an intricate and very specific order, which may be revised and redirected almost indefinitely. Such activity is subject to the conditioning and limitations arising from its sub-structure, but on its own level it may condition itself, having in view the achievement of chosen objectives.

Perhaps the best way further to understand the autonomous responsibility of which consciousness is capable is to consider those moments in which consciousness, so to speak, 'loses its grip' on past and future so that our conduct becomes less and less our own. This can happen in moments of extreme pain, fear, anger, acute embarrassment, or when we are subjected to a rigid routine of behavior like that imposed on a parade ground. Under such pressures, we lose sight of what the past might have taught us and we fail to entertain any detailed anticipation of the future. We are not then active beyond the prospect of a very narrow present, as in the experience of 'losing our temper.' When we regret our angry outburst we return toward a wider consciousness and 'become ourselves' once more. Thus when I ask forgiveness for impulsive thoughtlessness I may say 'I was not myself,' although of course I admit that I should have been. This admission stands at the threshold of moral philosophy. The selfhood of the individual, as it attains and surpasses the minimal continuity of personal identity, seems to have a normative or axiological dimension. We may come to recognize that to be oneself—that is, to possess a past and a future of one's own—is an achievement. It is this achievement that we may come most to respect in our evaluation of things human and it is toward the furthering of this achievement that our way of life should be oriented. This is the task of practical philosophy.

The physical and biological sub-structure of consciousness affords the background and the possibility of such a way of life. It is in this light that the philosopher should study them, if he is to contribute significant support to the basic ideology of our civilization. This task is not an easy one, and the student who accepts may often find himself discouraged. We must remember, however, that we have as much right to begin with the presence of consciousness in our world as with any other point of departure. By doing so we are more likely to keep in contact with the sense of value and moral common sense of our civilization. We should look forward to such a *rapprochement* and encourage its development in every way possible.

REFERENCES

[1] A. N. Whitehead, *Modes of Thought*, New York: Macmillan, 1938, p. 158.
[2] *Ibid.*, p. 228.

9

ON 'GOD' AND 'GOOD'

Iris Murdoch

I

To do philosophy is to explore one's own temperament, and yet at the same time one must attempt to discover the truth. It seems to me that there is a void in present-day moral philosophy. Areas peripheral to philosophy expand (psychology, political and social theory) or collapse (religion) without philosophy being able in the one case to encounter, and in the other case to rescue, the values involved. A working philosophical psychology is needed which can at least attempt to connect modern psychological terminology with a terminology concerned with virtue. We need a moral philosophy which can speak significantly of Freud and Marx, and out of which aesthetic and political views can be generated. We need a moral philosophy in which the concept of love, so rarely mentioned now by philosophers, can once again be made central.

It will be said, we have got a working philosophy, and one which is the proper heir to the past of European philosophy: existentialism. This philosophy does so far pervade the scene that philosophers—many linguistic analysts, for instance—who would not claim the name, do in fact work with, are imprisoned inside, existentialist concepts. I shall argue that existentialism is not, and cannot by tinkering be made, the philosophy we need. Although it is indeed the heir of the past, it is (it seems to me) an unrealistic and over-optimistic doctrine and the purveyor of certain false values. This is more obviously true of flimsier creeds, such as utilitarianism or 'humanism,' with which people might now attempt to fill the philosophical void.

The great merit of existentialism is that it at least professes and tries to be a philosophy one could live by. Kierkegaard described the Hegelian system as a grand palace set up by someone who then lived in a hovel or at best in the porter's lodge. A moral philosophy should be inhabited. Existentialism has shown itself capable of becoming a popular philosophy and of getting into the minds of those (e.g. Oxford philosophers) who have not sought it and may even be unconscious of its presence. However, although it can certainly inspire action, it seems to me to do so by a sort of romantic provocation rather than by its truth; and its pointers are often pointing in the wrong direction. Wittgenstein claimed that he brought the Cartesian era in philosophy to an end. Moral philosophy of an existentialist type (and almost all Western philosophy is of an existentialist type) is still Cartesian and egocentric. Briefly put, our picture of ourselves has become too grand, we have isolated, and identified ourselves with, an unrealistic conception of will, we have lost the vision of a reality separate from ourselves, and we have no adequate conception of original sin. Kierkegaard rightly observed that 'an ethic which ignores sin is an altogether useless science,' although he also added, 'but if it recognizes sin it is *eo ipso* beyond its sphere.'

Kant believed in Reason and Hegel believed in History, and for both this was a form of a belief in an external reality. Modern thinkers who believe in neither, but who remain within the tradition, are left with a denuded self whose only virtues are freedom, or at best sincerity, or, in the case of the British philosophers, an everyday reasonableness. Philosophy, on its other fronts, has been busy dismantling the old substantial picture of the 'self,' and ethics has not proved able to rethink this concept for moral purposes. The moral agent then is pictured as an isolated principle of will, or burrowing pinpoint of consciousness, inside, or beside, a lump of being which has been handed over to other disciplines, such as psychology or sociology. On the one hand a Luciferian philosophy of adventures of the will, and on the other natural science. Moral philosophy, and indeed morals, are thus undefended against an irresponsible and undirected self-assertion which goes easily hand in hand with some brand of pseudo-scientific determinism. An unexamined sense of the strength of the machine is combined with an illusion of leaping out of it. Sartre at his worst, and many British moral philosophers, represent this last dry distilment of Kant's view of the world. The study of motivation is surrendered to empirical science: will takes the place of the complex of motives and also of the complex of virtues.

The history of British philosophy since Moore represents intensively in miniature the special dilemmas of modern ethics. Empiricism, especially in the form given to it by Russell, and later by Wittgenstein, thrust ethics almost out of philosophy. Moral judgments were not factual, or truthful, and had no place in the world of the *Tractatus*. Moore, although he himself held a curious metaphysic of 'moral facts,' set the tone when he told us that we must carefully distinguish the question 'What things are good?' from the question 'What does "good" mean?' The answer to the latter question concerned the will. Good was indefinable (naturalism was a fallacy) because any offered good could be scrutinized by any individual by a 'stepping back' movement. This form of Kantianism still holds the field. Wittgenstein had attacked the idea of the Cartesian ego or substantial self and Ryle and others had developed the attack. A study of 'ordinary language' claimed (often rightly) to solve piecemeal problems in epistemology which had formerly been discussed in terms of the activities or faculties of a 'self.' (See John Austin's, it seems to me conclusive, book on certain problems of perception, *Sense and Sensibilia*.)

Ethics took its place in this scene. After puerile attempts to classify moral statements as exclamations or expressions of emotion, a more sophisticated neo-Kantianism with a utilitarian atmosphere has been developed. The idea of the agent as a privileged centre of will (for ever capable of 'stepping back') is retained, but, since the old-fashioned 'self' no longer clothes him he appears as an isolated will operating with the concepts of 'ordinary language,' so far as the field of morals is concerned. (It is interesting that although Wittgenstein's work has suggested this picture to others, he himself never used it.) Thus the will, and the psyche as an object of science, are isolated from each other and from the rest of philosophy. The cult of ordinary language goes with the claim to be neutral. Previous moral philosophers told us what we ought to do, that is they tried to answer both of Moore's questions. Linguistic analysis claims simply to give a philosophical description of the human pheno- menon of morality, without making any moral judgements. In fact the resulting picture of human conduct has a clear moral bias. The merits of linguistic analytical man are freedom (in the sense of detachment, rationality), responsibility, self-awareness, sincerity, and a lot of utilitarian common sense. There is of course no mention of sin, and no mention of love. Marxism is ignored, and there is on the whole no attempt at a *rapprochement* with psychology, although Professor Hampshire does try (see his Ernest Jones lecture) to develop the idea of self-awareness toward an ideal end-point by conceiving of 'the perfect psychoanalysis' which would make us perfectly self-aware and so perfectly detached and free.

Linguistic analysis of course poses for ethics the question of its relation with metaphysics. Can ethics be a form of empiricism? Many philosophers in the Oxford and Cambridge tradition would say yes. It is certainly a great merit of this tradition, and one which I would not wish to lose sight of, that it attacks every form of spurious unity. It is the traditional inspiration of the philosopher, but also his traditional vice, to believe that all is one. Wittgenstein says 'Let's see.' Sometimes problems turn out to be quite unconnected with each other, and demand types of solution which are not themselves closely related in any system. Perhaps it is a matter of temperament whether or not one is convinced that all is one. (My own temperament inclines to monism.) But let us postpone the question of whether, if we reject the relaxed empirical ethics of the British tradition (a cheerful amalgam of Hume, Kant, and Mill), and if we reject, too, the more formal existentialist systems, we wish to replace these with something which would have to be called a metaphysical theory. Let me now simply suggest ways in which I take the prevalent and popular picture to be unrealistic. In doing this, and throughout, my debt to Simone Weil will become evident.

Much of contemporary moral philosophy appears both unambitious and optimistic. Unambitious optimism is of course part of the Anglo-Saxon tradition; and it is also not surprising that a philosophy which analyses moral concepts on the basis of ordinary language should present a relaxed picture of a mediocre achievement. I think the charge is also true, though contrary to some appearances, of existentialism. An authentic mode of existence is presented as attainable by intelligence and force of will. The atmosphere is invigorating and tends to produce self-satisfaction in the reader, who feels himself to be a member of the élite, addressed by another one. Contempt for the ordinary human condition, together with a conviction of personal salvation, saves the writer from real pessimism. His gloom is superficial and conceals elation. (I think this to be true in different ways of both Sartre and Heidegger, though I am never too sure of having understood the latter.) Such attitudes contrast with the vanishing images of Christian theology which represented goodness as almost impossibly difficult, and sin as almost insuperable and certainly as a universal condition.

Yet modern psychology has provided us with what might be called a doctrine of original sin, a doctrine which most philosophers either deny (Sartre), ignore (Oxford and Cambridge), or attempt to render innocuous (Hampshire). When I speak in this context of modern psychology I mean primarily the work of Freud. I am not a 'Freudian' and the truth of this or that particular view of Freud does not here concern me, but it seems clear that Freud made an important discovery about the human mind and that he remains still the greatest scientist in the field which he opened. One may say that what he presents us with is a realistic and detailed picture of the fallen man. If we take the general outline of this picture seriously, and at the same time wish to do moral philosophy, we shall have to revise the current conceptions of will and motive very considerably. What seems to me, for these purposes, true and important in Freudian theory is as follows. Freud takes a thoroughly pessimistic view of human nature. He sees the psyche as an egocentric system of quasi-mechanical energy, largely determined by its own individual history, whose natural attachments are sexual, ambiguous, and hard for the subject to understand or control. Introspection reveals only the deep tissue of ambivalent motive, and fantasy is a stronger force than reason. Objectivity and unselfishness are not natural to human beings.

Of course Freud is saying these things in the context of a scientific therapy which aims not at making people good but at making them workable. If a moral philosopher says such things he must justify them not with scientific arguments but with arguments appropriate to philosophy; and in fact if he does say such things he will not be saying anything very new, since partially similar views have been expressed before in philosophy, as far back as Plato. It is important to look at Freud and his successors because they can give us more information about a mechanism the general nature of which we may discern without the help of science; and also because the ignoring of psychology may be a source of confusion. Some philosophers (e.g. Sartre) regard traditional psychoanalytical theory as a form of determinism and are prepared to deny it at all levels, and philosophers who ignore it often do so as part of an easy surrender to science of aspects of the mind which ought to interest them. But determinism as a total philosophical theory is not the enemy. Determinism as a philosophical theory is quite unproven, and it can be argued that it is not possible in principle to translate propositions about men making decisions and formulating viewpoints into the neutral language of natural science. (See Hampshire's brief discussion of this point in his book, *The Freedom of the Individual.*) The problem is to accommodate inside moral philosophy, and suggest methods of dealing with the fact that so much of human conduct is moved by mechanical energy of an egocentric kind. In the moral life the enemy is the fat, relentless ego. Moral philosophy is properly, and in the past has sometimes been, the discussion of this ego and of the techniques (if any) for its defeat. In this respect moral philosophy has shared some aims with religion. To say this is of course also to deny that moral philosophy should aim at being neutral.

What is a good man like? How can we make ourselves morally better? *Can* we make ourselves morally better? These are questions the philosopher should try to answer. We realize on reflection that we know little about good men. There are men in history who are traditionally thought of as having been good (Christ, Socrates, certain saints), but if we try to contemplate these men we find that the information about them is scanty and vague, and that, their great moments apart, it is the simplicity and directness of their diction which chiefly colours our conception of them as good. And if we consider contemporary candidates for goodness, if we know of any, we are likely to find them obscure, or else on closer inspection full of frailty. Goodness appears to be both rare and hard to picture. It is perhaps most convincingly met with in simple people— inarticulate, unselfish mothers of large families—but these cases are also the least illuminating.

It is significant that the idea of goodness (and of virtue) has been largely superseded in Western moral philosophy by the idea of rightness, supported perhaps by some conception of sincerity. This is to some extent a natural outcome of the disappearance of a permanent background to human activity: a permanent background, whether provided by God, by Reason, by History, or by the self. The agent, thin as a needle, appears in the quick flash of the choosing will. Yet existentialism itself, certainly in its French and Anglo-Saxon varieties, has, with a certain honesty, made evident the paradoxes of its own assumptions. Sartre tells us that when we deliberate the die is already cast, and Oxford philosophy has developed no serious theory of motivation. The agent's freedom, indeed his moral quality, resides in his choices, and yet we are not told what prepares him for the choices. Sartre can admit, with bravado, that we choose out of some sort of pre-existent condition, which he also confusingly calls a choice, and Richard Hare declares that since the identification of mental data, such as 'intentions,' is so philosophically difficult we had better say that a man is morally the set of his actual choices. That visible motives do not necessitate acts is taken by Sartre as a cue for asserting an irresponsible freedom as an obscure postulate; that motives do not readily yield to 'introspection' is taken by the British philosophers as an excuse for forgetting them and talking about 'reasons' instead. These views seem both unhelpful to the moral pilgrim and also profoundly unrealistic. Moral choice is often a mysterious matter. Kant thought so, and he pictured the mystery in terms of an indiscernible balance between a pure rational agent and an impersonal mechanism, neither of which represented what we normally think of as personality; much existentialist philosophy is in this respect, though often covertly, Kantian. But should not the mystery of choice be conceived of in some other way?

We have learned from Freud to picture 'the mechanism' as something highly individual and personal, which is at the same time very powerful and not easily understood by its owner. The self of psychoanalysis is certainly substantial enough. The existentialist picture of choice, whether it be surrealist or rational, seems unrealistic, over-optimistic, romantic, because it ignores what appears at least to be a sort of continuous background with a life of its own; and it is surely in the tissue of that

239

life that the secrets of good and evil are to be found. Here neither the inspiring ideas of freedom, sincerity and fiats of will, nor the plain wholesome concept of a rational discernment of duty, seem complex enough to do justice to what we really are. What we really are seems much more like an obscure system of energy out of which choices and visible acts of will emerge at intervals in ways which are often unclear and often dependent on the condition of the system in between the moments of choice.

II

If this is so, one of the main problems of moral philosophy might be formulated thus: are there any techniques for the purification and reorientation of an energy which is naturally selfish, in such a way that when moments of choice arrive we shall be sure of acting rightly? We shall also have to ask whether, if there are such techniques, they should be simply described, in quasi-psychological terms, perhaps in psychological terms, or whether they can be spoken of in a more systematic philosophical way. I have already suggested that a pessimistic view which claims that goodness is the almost impossible countering of a powerful egocentric mechanism already exists in traditional philosophy and in theology. The technique which Plato thought appropriate to this situation I shall partly discuss later. Much closer and more familiar to us are the techniques of religion, of which the most widely practised is prayer. What becomes of such a technique in a world without God, and can it be transformed to supply at least part of the answer to our central question?

Prayer is properly not petition, but simply an attention to God which is a form of love. With it goes the idea of grace, of a supernatural assistance to human endeavour which overcomes empirical limitations of personality. What is this attention like, and can those who are not religious believers still conceive of profiting by such an activity? Let us pursue the matter by considering what the traditional object of this attention was like and by what means it affected its worshippers. I shall suggest that God was (or is) a *single perfect transcendent non-representable* and *necessarily real object of attention*; and I shall go on to suggest that moral philosophy should attempt to retain a central concept which has all these characteristics. I shall consider them one by one, although to a large extent they interpenetrate and overlap.

Let us take first the notion of an object of attention. The religious believer, especially if his God is conceived of as a person, is in the fortunate position of being able to focus his thought upon something which is a source of energy. Such focusing, with such results, is natural to human beings. Consider being in love. Consider too the attempt to check being in love, and the need in such a case of another object to attend to. Where strong emotions of sexual love, or of hatred, resentment, or jealousy are concerned, 'pure will' can usually achieve little. It is small use telling oneself 'Stop being in love, stop feeling resentment, be just.' What is needed is a reorientation which will provide an energy of a different kind, from a different source. Notice the metaphors of orientation and of looking. The neo-Kantian existentialist 'will' is a principle of pure movement. But how ill this describes what it is like for us to alter. Deliberately falling out of love is not a jump of the will, it is the acquiring of new objects of attention and thus of new energies as a result of refocusing. The metaphor of orientation may indeed also cover moments when recognizable 'efforts of will' are made, but explicit efforts of will are only a part of the whole situation. That God, attended to, is a powerful source of (often good) energy is a psychological fact. It is also a psychological fact, and one of importance in moral philosophy, that we can all receive moral help by focusing our attention upon things which are valuable: virtuous people, great art, perhaps (I will discuss this later) the idea of goodness itself. Human beings are naturally 'attached' and when an attachment seems painful or bad it is most readily displaced by another attachment, which an attempt at attention can encourage. There is nothing odd or mystical about this, nor about the fact that our ability to act well 'when the time comes' depends partly, perhaps largely, upon the quality of our habitual objects of attention. 'Whatsoever things are true, whatsoever things are honest, whatsoever things are just, whatsoever things are pure, whatsoever things are lovely, whatsoever things are of good report; if there be any virtue, and if there be any praise, think on these things.' The notion that value should be in some sense *unitary*, or even that there should be a single supreme value concept, may seem, if one surrenders the idea of God, far from obvious. Why should there not be many different kinds of independent moral values? Why should all be one here? The madhouses of the world are filled with people who are convinced that all is one. It might be said that 'all is one' is a dangerous falsehood at any level except the highest; and can that be discerned at all? That a belief in the unity, and also in the hierarchical order, of the moral world has a psychological importance is fairly evident. The notion that 'it all somehow must make sense,' or 'there is a best decision here,' preserves from despair: the difficulty is how to entertain this consoling notion in a way which is not false. As soon as any idea is a consolation the tendency to falsify it becomes strong: hence the traditional problem of preventing the idea of God from degenerating in the believer's mind. It is true that the intellect naturally seeks unity; and in the sciences, for instance, the assumption of unity consistently rewards the seeker. But how can this dangerous idea be used in morals? It is useless to ask 'ordinary language' for a judgment, since we are dealing with concepts which are not on display in ordinary language or unambiguously tied up to ordinary words. Ordinary language is not a philosopher.

We might, however, set out from an ordinary language situation by reflecting upon the virtues. The concepts of the virtues, and the familiar words which name them, are important, since they help to make certain potentially nebulous areas of experience more open to inspection. If we reflect upon the nature of the virtues we are constantly led to consider their relation to each other. The idea of an 'order' of virtues suggests itself, although it might of course be difficult to state this in any systematic form. For instance, if we reflect upon courage and ask why we think it to be a virtue, what kind of courage is the highest, what distinguishes courage from rashness, ferocity, self-assertion, and so on, we are bound, in our explanation, to use the names of other virtues. The best kind of courage (that which would make a man act unselfishly in a concentration camp) is steadfast, calm, temperate, intelligent, loving. . . . This may not in fact be exactly the right description, but it is the right sort of description. Whether there is a single supreme principle in the united world of the virtues, and whether the name of that

242

principle is love, is something which I shall discuss below. All I suggest here is that reflection rightly tends to unify the moral world, and that increasing moral sophistication reveals increasing unity. What is it like to be just? We come to understand this as we come to understand the relationship between justice and the other virtues. Such a reflection requires and generates a rich and diversified vocabulary for naming aspects of goodness. It is a shortcoming of much contemporary moral philosophy that it eschews discussion of the separate virtues, preferring to proceed directly to some sovereign concept such as sincerity, or authenticity, or freedom, thereby imposing, it seems to me, an unexamined and empty idea of unity, and impoverishing our moral language in an important area.

We have spoken of an 'object of attention' and of an unavoidable sense of 'unity.' Let us now go on to consider, thirdly, the much more difficult idea of 'transcendence.' All that has been said so far could be said without benefit of metaphysics. But now it may be asked: are you speaking of a transcendent authority or of a psychological device? It seems to me that the idea of the transcendent, in some form or other, belongs to morality: but it is not easy to interpret. As with so many of these large elusive ideas it readily takes on forms which are false ones. There is a false transcendence, as there is a false unity, which is generated by modern empiricism: a transcendence which is in effect simply an exclusion, a relegation of the moral to a shadowy existence in terms of emotive language, imperatives, behaviour patterns, attitudes. 'Value' does not belong inside the world of truth functions, the world of science and factual propositions. So it must live somewhere else. It is then attached somehow to the human will, a shadow clinging to a shadow. The result is the sort of dreary moral solipsism which so many so-called books on ethics now purvey. An instrument for criticizing the false transcendence, in many of its forms, has been given to us by Marx in the concept of alienation. Is there, however, any true transcendence, or is this idea always a consoling dream projected by human need on to an empty sky?

It is difficult to be exact here. I would like to say something like this. One might start from the assertion that morality, goodness, is a form of realism. The idea of a really good man living in a private dream world seems unacceptable. Of course a good man may be infinitely eccentric, but he must know certain things about his surroundings, most obviously the existence of other people and their claims. The chief enemy of excellence in morality (and also in art) is personal fantasy: the tissue of self-aggrandizing and consoling wishes and dreams which prevents one from seeing what is there outside one. Rilke said of Cézanne that he did not paint 'I like it,' he painted 'There it is.' This is not easy, and requires, in art or morals, a discipline. One might say here that art is an excellent analogy of morals, or indeed that it is in this respect a case of morals. We cease to be in order to attend to the existence of something else, a natural object, a person in need. We can see in mediocre art, where perhaps it is even more clearly seen than in mediocre conduct, the intrusion of fantasy, the assertion of self, the dimming of any reflection of the real world.

It may be agreed that the direction of attention should properly be outward, away from self, but it will be said that it is a long step from the idea of realism to the idea of transcendence. I think, however, that these two ideas are related, and one can see their relation particularly in the case of our apprehension of beauty. The link here is the concept of indestructibility or incorruptibility. What is truly beautiful is 'inaccessible' and cannot be possessed or destroyed. The statue is broken, the flower fades, the experience ceases, but something has not suffered from decay and mortality. Almost anything that consoles us is a fake, and it is not easy to prevent this idea from degenerating into a vague Shelleyan mysticism. In the case of the idea of a transcendent personal God the degeneration of the idea seems scarcely avoidable: theologians are busy at their desks at this very moment trying to undo the results of this degeneration. In the case of beauty, whether in art or in nature, the sense of separateness from the temporal process is connected perhaps with concepts of perfection of form and 'authority' which are not easy to transfer into the field of morals. Here I am not sure if this is an analogy or an instance. It is as if we can see beauty itself in a way in which we cannot see goodness itself. (Plato says this at *Phaedrus* 250E.) I can *experience* the transcendence of the beautiful, but (I think) not the transcendence of the good. Beautiful things contain beauty in a way in which good acts do not exactly contain good, because beauty is partly a matter of the senses. So if we speak of good as transcendent we are speaking of something rather more complicated and which cannot be experienced, even when we see the unselfish man in the concentration camp. One might be tempted to use the word 'faith' here if it could be purged of its religious associations. 'What is truly good is incorruptible and indestructible.' 'Goodness is not in this world.' These sound like highly metaphysical statements. Can we give them any clear meaning or are they just things one 'feels inclined to say'?

I think the idea of transcendence here connects with two separate ideas, both of which I will be further concerned with later: *perfection* and *certainty*. Are we not certain that there is a 'true direction' toward better conduct, that goodness 'really matters,' and does not that certainty about a standard suggest an idea of permanance which cannot be reduced to psychological or any other set of empirical terms? It is true, and this connects with considerations already put forward under the heading of 'attention,' that there is a psychological power which derives from the mere idea of a transcendent object, and one might say further from a transcendent object which is to some extent mysterious. But a reductive analysis in, for instance, Freudian terms, or Marxist terms, seems properly to apply here only to a degenerate form of a conception about which one remains certain that a higher and invulnerable form must exist. The idea admittedly remains very difficult. How is one to connect the realism which must involve a clear-eyed contemplation of the misery and evil of the world with a sense of an uncorrupted good without the latter idea becoming the merest consolatory dream? (I think this puts a central problem in moral philosophy.) Also, what is it for someone, who is not a religious believer and not some sort of mystic, to apprehend some separate 'form' of goodness behind the multifarious cases of good behavior? Should not this idea be reduced to the much more intelligible notion of the interrelation of the virtues, plus a purely subjective sense of the certainty of judgments?

At this point the hope of answering these questions might lead us on to consider the next, and closely related, 'attributes': *perfection* (absolute good) and *necessary existence*. These attributes are indeed so closely connected that from some points of view they are the same. (Ontological proof.) It may seem curious to wonder whether the idea of perfection (as opposed to the idea of merit or improvement) is really an important one, and what sort of role it can play. Well, is it important to measure and compare things and know just how good they are? In any field which interests or concerns us I think we would say yes. A deep understanding of any field of human activity (painting, for instance) involves an increasing revelation of degrees of excellence and often a revelation of there being in fact little that is very good and nothing that is perfect. Increasing understanding of human conduct operates in a similar way. We come to perceive scales, distances, standards, and may incline to see as less than excellent what previously we were prepared to 'let by.' (This need not of course hinder the operation of the virtue of tolerance: tolerance can be, indeed ought to be, clear-sighted.) The idea of perfection works thus within a field of study, producing an increasing sense of direction. To say this is not perhaps to say anything very startling; and a reductionist might argue that an increasingly refined ability to compare need not imply anything beyond itself. The idea of perfection might be, as it were, empty.

Let us consider the case of conduct. What of the command 'Be ye therefore perfect'? Would it not be more sensible to say 'Be ye therefore slightly improved'? Some psychologists warn us that if our standards are too high we shall become neurotic. It seems to me that the idea of love arises necessarily in this context. The idea of perfection moves, and possibly changes, us (as artist, worker, agent) because it inspires love in the part of us that is most worthy. One cannot feel unmixed love for a mediocre moral standard any more than one can for the work of a mediocre artist. The idea of perfection is also a natural producer of order. In its *light* we come to see that A, which superficially resembles B, is really better than B. And this can occur, indeed must occur, without our having the sovereign idea in any sense 'taped.' In fact it is in its nature that we cannot get it taped. This is the true sense of the 'indefinability' of the good, which was given a vulgar sense by Moore and his followers. It lies always beyond, and it is from this beyond that it exercises its *authority*. Here again the word seems naturally in place, and it is in the work of artists that we see the operation most clearly. The true artist is obedient to a conception of perfection to which his work is constantly related and re-related in what seems an external manner. One may of course try to 'incarnate' the idea of perfection by saying to oneself 'I want to write like Shakespeare' or 'I want to paint like Piero.' But of course one knows that Shakespeare and Piero, though almost gods, are not gods, and that one has got to do the thing oneself alone and differently, and that beyond the details of craft and criticism there is only the magnetic non-representable idea of the good which remains not 'empty' so much as mysterious. And thus too in the sphere of human conduct.

It will be said perhaps: are these not simply empirical generalizations about the psychology of effort or improvement or what status do you wish them to have? Is it just a matter of 'this works' or 'it is as if this were so'? I shall discuss this more fully later. But let us consider what, if our subject of discussion were not Good but God, the reply might be. God exists *necessarily*. Everything else which exists exists contingently. What can this mean? I am assuming that there is no plausible 'proof' of the existence of God except some form of the ontological proof, a 'proof' incidentally which must now take on an increased importance in theology as a result of the recent 'demythologizing.' If considered carefully, however, the ontological proof is seen to be not exactly a proof but rather a clear assertion of faith (it is often admitted to be appropriate only for those already convinced), which could only confidently be made on the basis of a certain amount of experience. This assertion could be put in various ways. The desire for God is certain to receive a response. My conception of God contains the certainty of its own reality. God is an object of love which uniquely excludes doubt and relativism. Such obscure statements would of course receive little sympathy from analytical philosophers, who would divide their content between psychological fact and metaphysical nonsense, and who might remark that one could just as well take 'I know that my Redeemer liveth,' as asserted by Handel, as a philosophical argument. Whether they are right about 'God' I leave aside: but what about the fate of 'Good'? The difficulties seem similar. What status can we give to the idea of certainty which does seem to attach itself to the idea of good? Or to the notion that we must receive a return when good is sincerely desired? (The concept of grace can be readily

247

secularized.) What is formulated here seems unlike an 'as if' or a 'it works.' Of course one must avoid here, as in the case of God, any heavy material connotation of the misleading word 'exist.' Equally, however, a purely subjective conviction of certainty, which could receive a ready psychological explanation, seems less than enough. Could the problem really be subdivided without residue by a careful linguistic analyst into parts which he would deem innocuous?

A little light may be thrown on the matter if we return now, after the intervening discussion, to the idea of *'realism'* which was used earlier in a normative sense: that is, it was assumed that it was better to know what was real than to be in a state of fantasy or illusion. It is true that human beings cannot bear much reality; and a consideration of what the effort to face reality is like, and what are its techniques, may serve both to illuminate the necessity or certainty which seems to attach to 'the Good,' and also to lead on to a reinterpretation of 'will' and 'freedom' in relation to the concept of love. Here again it seems to me that art is the clue. Art presents the most comprehensible examples of the almost irresistible human tendency to seek consolation in fantasy and also of the effort to resist this and the vision of reality which comes with success. Success in fact is rare. Almost all art is a form of fantasy-consolation and few artists achieve the vision of the real. The talent of the artist can be readily, and is naturally, employed to produce a picture whose purpose is the consolation and aggrandisement of its author and the projection of his personal obsessions and wishes. To silence and expel self, to contemplate and delineate nature with a clear eye, is not easy and demands a moral discipline. A great artist is, in respect of his work, a good man, and, in the true sense, a free man. The consumer of art has an analogous task to its producer: to be disciplined enough to see as much reality in the work as the artist has succeeded in putting into it, and not to 'use it as magic.' The appreciation of beauty in art or nature is not only (for all its difficulties) the easiest available spiritual exercise; it is also a completely adequate entry into (and not just analogy of) the good life, since it *is* the checking of selfishness in the interest of seeing the real. Of course great artists are 'personalities' and have special styles; even Shakespeare occasionally, though very occasionally, reveals a personal obsession. But the greatest art is 'impersonal' because it shows us the world, our world and not another one, with a clarity which startles and delights us simply because we are not used to looking at the real world at all. Of course, too, artists are pattern-makers. The claims of form and the question of 'how much form' to impose constitutes one of the chief problems of art. But it is when form is used to isolate, to explore, to display something which is true that we are most highly moved and enlightened. Plato says (*Republic*, VII, 532) that the *technai* have the power to lead the best part of the soul to the view of what is most excellent in reality. This well describes the role of great art as an educator and revealer. Consider what we learn from contemplating the characters of Shakespeare or Tolstoy or the paintings of Velasquez or Rembrandt. What is learnt here is something about the real quality of human nature, when it is envisaged, in the artist's just and compassionate vision, with a clarity which does not belong to the self-centred rush of ordinary life.

It is important too that great art teaches us how real things can be looked at and loved without being seized and used, without being appropriated into the greedy organism of the self. This exercise of *detachment* is difficult and valuable whether the thing contemplated is a human being or the root of a tree or the vibration of a colour or a sound. Unsentimental contemplation of nature exhibits the same quality of detachment: selfish concerns vanish, nothing exists except the things which are seen. Beauty is that which attracts this particular sort of unselfish attention. It is obvious here what is the role, for the artist or spectator, of exactness and good vision: unsentimental, detached, unselfish, objective attention. It is also clear that in moral situations a similar exactness is called for. I would suggest that the authority of the Good seems to us something necessary because the realism (ability to perceive reality) required for goodness is a kind of intellectual ability to perceive what is true, which is automatically at the same time a suppression of self. *The necessity of the good is then an aspect of the kind of necessity involved in any technique for exhibiting fact.* In thus treating realism, whether of artist or of agent, as a moral achievement, there is of course a further assumption to be made in the field of morals: that true vision occasions right conduct. This could be uttered simply as an enlightening tautology:

but I think it can in fact be supported by appeals to experience. The more the separateness and differentness of other people is realized, and the fact seen that another man has needs and wishes as demanding as one's own, the harder it becomes to treat a person as a thing. That it is realism which makes great art great remains too as a kind of proof.

If, still led by the clue of art, we ask further questions about the faculty which is supposed to relate us to what is real and thus bring us to what is good, the idea of compassion or love will be naturally suggested. It is not simply that suppression of self is required before accurate vision can be obtained. The great artist sees his objects (and this is true whether they are sad, absurd, repulsive or even evil) in a light of justice and mercy. The direction of attention is, contrary to nature, outward, away from self which reduces all to a false unity, toward the great surprising variety of the world, and the ability so to direct attention is love.

III

One might at this point pause and consider the picture of human personality, or the soul, which has been emerging. It is in the capacity to love, that is to *see*, that the liberation of the soul from fantasy consists. The freedom which is a proper human goal is the freedom from fantasy, that is the realism of compassion. What I have called fantasy, the proliferation of blinding self-centred aims and images, is itself a powerful system of energy, and most of what is often called 'will' or 'willing' belongs to this system. What counteracts the system is attention to reality inspired by, consisting of, love. In the case of art and nature such attention is immediately rewarded by the enjoyment of beauty. In the case of morality, although there are sometimes rewards, the idea of a reward is out of place. Freedom is not strictly the exercise of the will, but rather the experience of accurate vision which, when this becomes appropriate, occasions action. It is what lies behind and in between actions and prompts them that is important, and it is this area which should be purified. By the time the moment of choice has arrived the quality of attention has probably determined the nature of the act. This fact produces that curious separation between consciously rehearsed motives and action which is sometimes wrongly taken as an experience of freedom. (*Angst.*) Of course this is not to say that good 'efforts of will' are always useless or always fakes. Explicit and immediate 'willing' can play some part, especially as an inhibiting factor. (The daemon of Socrates only told him what not to do.)

In such a picture sincerity and self-knowledge, those popular merits, seem less important. It is an attachment to what lies outside the fantasy mechanism, and not a scrutiny of the mechanism itself, that liberates. Close scrutiny of the mechanism often merely strengthens its power. 'Self-knowledge,' in the sense of a minute understanding of one's own machinery, seems to me, except at a fairly simple level, usually a delusion. A sense of such self-knowledge may of course be induced in analysis for therapeutic reasons, but 'the cure' does not prove the alleged knowledge genuine. Self is as hard to see justly as other things, and when clear vision has been achieved, self is a correspondingly smaller and less interesting object. A chief enemy to such clarity of vision, whether in art or morals, is the system to which the technical name of sado-masochism has been given. It is the peculiar subtlety of this system that, while constantly leading attention and energy back into the self, it can produce, almost all the way as it were to the summit, plausible imitations of what is good. Refined sado-masochism can ruin art which is too good to be ruined by the cruder vulgarities of self-indulgence. One's self is interesting, so one's motives are interesting, and the unworthiness of one's motives is interesting. Fascinating too is the alleged relation of master to slave, of the good self to the bad self which, oddly enough, ends in such curious compromises. (Kafka's struggle with the devil which ends up in bed.) The bad self is prepared to suffer but not to obey until the two selves are friends and obedience has become reasonably easy or at least amusing. In reality the good self is very small indeed, and most of what appears good is not. The truly good is not a friendly tyrant to the bad, it is its deadly foe. Even suffering itself can play a demonic role here, and the ideas of guilt and punishment can be the most subtle tool of the ingenious self. The idea of suffering confuses the mind and in certain contexts (the context of 'sincere self-examination,' for instance) can masquerade as a purification. It is rarely this, for unless it is very intense indeed it is far too interesting. Plato does not say that philosophy is the study of suffering, he says it is the study of death (*Phaedo* 64A), and these ideas are totally dissimilar. That moral improvement involves suffering is usually true; but the suffering is the by-product of a new orientation and not in any sense an end in itself.

I have spoken of the real which is the proper object of love, and of knowledge which is freedom. The word 'good' which has been moving about in the discussion should now be more explicitly considered. Can good itself be in any sense 'an object of attention'? And how does this problem relate to 'love of the real'? Is there, as it were, a substitute for prayer, that most profound and effective of religious techniques? If the energy and violence of will, exerted on occasions of choice, seems less important than the quality of attention which determines our real attachments, how do we alter and purify that attention and make it more realistic? Is the *via negativa* of the will, its occasional ability to stop a bad move, the only or most considerable conscious power which we can exert? I think there is something analogous to prayer, though it is something difficult to describe, and which the higher subtleties of the self can often falsify; I am not here thinking of any quasi-religious meditative technique, but of something which belongs to the moral life of the ordinary person. The idea of contemplation is hard to understand and maintain in a world increasingly without sacraments and ritual and in which philosophy has (in many respects rightly) destroyed the old substantial conception of the self. A sacrament provides an external visible place for an internal invisible act of the spirit. Perhaps one needs too an analogy of the concept of the sacrament, though this must be treated with great caution. Behaviouristic ethics denies the importance, because it questions the identity, of anything prior to or apart from action which decisively occurs 'in the mind.' The apprehension of beauty, in art or in nature, often in fact seems to us like a temporally located spiritual experience which is a source of good energy. It is not easy, however, to extend the idea of such an influential experience to occasions of thinking about people or action, since clarity of thought and purity of attention become harder and more ambiguous when the object of attention is something moral.

It is here that it seems to me to be important to retain the idea of Good as a central point of reflection, and here too we may see the significance of its indefinable and non-representable character. Good, not will, is transcendent. Will is the natural energy of the psyche which is sometimes employable for a worthy purpose. Good is the focus of attention when an intent to be virtuous co-exists (as perhaps it almost always does) with some unclarity of vision. Here, as I have said earlier, beauty appears as the visible and accessible aspect of the Good. The Good itself is not visible. Plato pictured the good man as eventually able to look at the sun. I have never been sure what to make of this part of the myth. While it seems proper to represent the Good as a centre or focus of attention, yet it cannot quite be thought of as a 'visible' one in that it cannot be experienced or represented or defined. We can certainly know more or less where the sun is; it is not so easy to imagine what it would be like to look at it. Perhaps indeed only the good man knows what this is like; or perhaps to look at the sun is to be gloriously dazzled and to see nothing. What does seem to make perfect sense in the Platonic myth is the idea of the Good as the source of light which reveals to us all things as they really are. All just vision, even in the strictest problems of the intellect, and *a fortiori* when suffering or wickedness have to be perceived, is a moral matter. The same virtues, in the end the same virtue (love), are required throughout, and fantasy (self) can prevent us from seeing a blade of grass just as it can prevent us from seeing another person. An increasing awareness of 'goods' and the attempt (usually only partially successful) to attend to them purely, without self, brings with it an increasing awareness of the unity and interdependence of the moral world. One-seeking intelligence is the image of 'faith.' Consider what it is like to increase one's understanding of a great work of art.

I think it is more than a verbal point to say that what should be aimed at is goodness, and not freedom or right action, although right action, and freedom in the sense of humility, are the natural products of attention to the Good. Of course right action is important in itself, with an importance which is not difficult to understand. But it should provide the starting-point of reflection and not its conclusion. Right action, together with the steady extension of the area of strict obligation, is a proper criterion of virtue. Action also tends to confirm, for better or worse, the background of attachment from which it issues.

Action is an occasion for grace, or for its opposite. However, the aim of morality cannot be simply action. Without some more positive conception of the soul as a substantial and continually developing mechanism of attachments, the purification and re-orientation of which must be the task of morals, 'freedom' is readily corrupted into self-assertion and 'right action' into some sort of *ad hoc* utilitarianism. If a scientifically minded empiricism is not to swallow up the study of ethics completely, philosophers must try to invent a terminology which shows how our natural psychology can be altered by conceptions which lie beyond its range. It seems to me that the Platonic metaphor of the idea of the Good provides a suitable picture here. With this picture must of course be joined a realistic conception of natural psychology (about which almost all philosophers seem to me to have been too optimistic) and also an acceptance of the utter lack of finality in human life. The Good has nothing to do with purpose, indeed it excludes the idea of purpose. 'All is vanity' is the beginning and the end of ethics. The only genuine way to be good is to be good 'for nothing' in the midst of a scene where every 'natural' thing, including one's own mind, is subject to chance, that is, to necessity. That 'for nothing' is indeed the experienced correlate of the invisibility or non-representable blankness of the idea of Good itself.

IV

I have suggested that moral philosophy needs a new and, to my mind, more realistic, less romantic, terminology if it is to rescue thought about human destiny from a scientifically minded empiricism which is not equipped to deal with the real problems. Linguistic philosophy has already begun to join hands with such an empiricism, and most existentialist thinking seems to me either optimistic romancing or else something positively Lucifer-ian. (Possibly Heidegger is Lucifer in person.) However, at this point someone might say, all this is very well, the only difficulty is that none of it is true. Perhaps indeed all is vanity, *all* is vanity, and there is no respectable intellectual way of protecting people from despair. The world just is hopelessly evil and should you, who speak of realism, not go all the way toward being realistic about this? To speak of Good in this portentous manner

254

is simply to speak of the old concept of God in a thin disguise. But at least 'God' could play a real consoling and encouraging role. It makes sense to speak of loving God, a person, but very little sense to speak of loving Good, a concept. 'Good' even as a fiction is not likely to inspire, or even be comprehensible to, more than a small number of mystically minded people who, being reluctant to surrender 'God,' fake up 'Good' in his image, so as to preserve some kind of hope. The picture is not only purely imaginary, it is not even likely to be effective. It is very much better to rely on simple popular utilitarian and existentialist ideas, together with a little empirical psychology, and perhaps some doctored Marxism, to keep the human race going. Day-to-day empirical common sense must have the last word. All specialized ethical vocabularies are false. The old serious metaphysical quest had better now be let go, together with the outdated concept of God the Father.

I am often more than half persuaded to think in these terms myself. It is frequently difficult in philosophy to tell whether one is saying something reasonably public and objective, or whether one is merely erecting a barrier, special to one's own temperament, against one's own personal fears. (It is always a significant question to ask about any philosopher: what is he afraid of?) Of course one is afraid that the attempt to be good may turn out to be meaningless, or at best something vague and not very important, or turn out to be as Nietzsche described it, or that the greatness of great art may be an ephemeral illusion. Of the 'status' of my arguments I will speak briefly below. That a glance at the scene prompts despair is certainly the case. The difficulty indeed is to look at all. If one does not believe in a personal God there is no 'problem' of evil, but there is the almost insuperable difficulty of looking properly at evil and human suffering. It is very difficult to concentrate attention upon suffering and sin, in others or in oneself, without falsifying the picture in some way while making it bearable. (For instance, by the sado-masochistic devices I mentioned earlier.) Only the very greatest art can manage it, and that is the only public evidence that it can be done at all. Kant's notion of the sublime, though extremely interesting, possibly even more interesting than Kant realized, is a kind of romanticism. The spectacle of huge and appalling things can indeed exhilarate, but

255

usually in a way that is less than excellent. Much existentialist thought relies upon such a 'thinking reed' reaction
which is nothing more than a form of romantic self-assertion.
It is not this which will lead a man on to unselfish behavior
in the concentration camp. There is, however, something in the
serious attempt to look compassionately at human things which
automatically suggests that 'there is more than this.' The
'there is more than this,' if it is not to be corrupted by some
sort of quasi-theological finality, must remain a very tiny spark
of insight, something with, as it were, a metaphysical position
but no metaphysical form. But it seems to me that the spark is
real, and that great art is evidence of its reality. Art indeed,
so far from being a playful diversion of the human race, is the
place of its most fundamental insight, and the centre to which
the more uncertain steps of metaphysics must constantly return.

As for the élite of mystics, I would say no to the term 'élite.'
Of course philosophy has its own terminology, but what it
attempts to describe need not be, and I think is not in this case,
removed from ordinary life. Morality has always been connected
with religion and religion with mysticism. The disappearance of
the middle term leaves morality in a situation which is certainly
more difficult but essentially the same. The background to
morals is properly some sort of mysticism, if by this is meant a
non-dogmatic essentially unformulated faith in the reality of the
Good, occasionally connected with experience. The virtuous
peasant knows, and I believe he will go on knowing, in spite of
the removal or modification of the theological apparatus,
although what he knows he might be at a loss to say. This view
is of course not amenable even to a persuasive philosophical
proof and can easily be challenged on all sorts of empirical
grounds. However, I do not think that the virtuous peasant
will be without resources. Traditional Christian superstition
has been compatible with every sort of conduct, from bad to
good. There will doubtless be new superstitions; and it will
remain the case that some people will manage effectively to love
their neighbours. I think the 'machinery of salvation' (if it exists)
is essentially the same for all. There is no complicated secret
doctrine. We are all capable of criticizing, modifying and extending the area of strict obligation which we have inherited. Good
is non-representable and indefinable. We are all mortal and

equally at the mercy of necessity and chance. These are the true respects in which all men are brothers.

On the status of the argument there is perhaps little, or else too much, to say. In so far as there is an argument it has already, in a compressed way, occurred. Philosophical argument is almost always inconclusive, and this one is not of the most rigorous kind. This is not a sort of pragmatism or a philosophy of 'as if.' If someone says, 'Do you then believe that the Idea of the Good exists?' I reply, 'No, not as people used to think God existed.' All one can do is to appeal to certain areas of experience, pointing out certain features, and using suitable metaphors and inventing suitable concepts where necessary to make these features visible. No more, and no less, than this is done by the most empirically minded of linguistic philosophers. As there is no philosophical or scientific proof of total determinism the notion is at least allowable that there is a part of the soul which is free from the mechanism of empirical psychology. I would wish to combine the assertion of such a freedom with a strict and largely empirical view of the mechanism itself. Of the very small area of 'freedom,' that in us which attends to the real and is attracted by the good, I would wish to give an equally rigorous and perhaps pessimistic account.

I have not spoken of the role of love in its everyday manifestations. If one is going to speak of great art as 'evidence,' is not ordinary human love an even more striking evidence of a transcendent principle of good? Plato was prepared to take it as a starting-point. (There are several starting-points.) One cannot but agree that in some sense this is the most important thing of all; and yet human love is normally too profoundly possessive and also too 'mechanical' to be a place of vision. There is a paradox here about the nature of love itself. That the highest love is in some sense impersonal is something which we can indeed see in art, but which I think we cannot see clearly, except in a very piecemeal manner, in the relationships of human beings. Once again the place of art is unique. The image of the Good as a transcendent magnetic centre seems to me the least corruptible and most realistic picture for us to use in our reflections upon the moral life. Here the philosophical 'proof,' if there is one, is the same as the moral 'proof.' I would rely especially upon arguments from experience concerned with the realism which we perceive to be connected with goodness, and with the love and detachment which is exhibited in great art.

I have throughout this paper assumed that 'there is no God' and that the influence of religion is waning rapidly. Both these assumptions may be challenged. What seems beyond doubt is that moral philosophy is daunted and confused, and in many quarters discredited and regarded as unnecessary. The vanishing of the philosophical self, together with the confident filling in of the scientific self, has led in ethics to an inflated and yet empty conception of the will, and it is this that I have been chiefly attacking. I am not sure how far my positive suggestions make sense. The search for unity is deeply natural, but like so many things which are deeply natural may be capable of producing nothing but a variety of illusions. What I feel sure of is the inadequacy, indeed the inaccuracy, of utilitarianism, linguistic behaviourism, and current existentialism in any of the forms with which I am familiar. I also feel sure that moral philosophy ought to be defended and kept in existence as a pure activity, or fertile area, analogous in importance to unapplied mathematics or pure 'useless' historical research. Ethical theory has affected society, and has reached as far as to the ordinary man, in the past, and there is no good reason to think that it cannot do so in the future. For both the collective and the individual salvation of the human race art is doubtless more important than philosophy, and literature most important of all. But there can be no substitute for pure, disciplined, professional speculation: and it is from these two areas, art and ethics, that we must hope to generate concepts worthy, and also able, to guide and check the increasing power of science.

10

THREE DIRECTIONS
OF PHENOMENOLOGY

William H. Bossart

Since Descartes, philosophers have tended to accept the view that philosophy must begin by examining the knowing mind. Only through an analysis of the powers of the mind can we hope to determine the nature and limits of human knowledge. And only if our knowledge is founded on the immediate data of consciousness can it claim to be certain. Yet once it withdraws to the confines of consciousness, philosophy is confronted by the difficulties of establishing a relation between mind and body and the existence of the external world. Hence philosophers have become increasingly suspicious of Descartes' point of departure. Descartes, we are told, committed an illegitimate bifurcation of nature. Instead of beginning with a disembodied consciousness, philosophy must begin with the union of body and mind and with the fact that man is immediately given to himself as in a world.

But the mere *assertion* that we cannot doubt what Descartes did put in doubt does not relieve us of these difficulties. It seems to me that Descartes was correct in holding that philosophy must begin with a sphere of experience which in some way justifies itself. Otherwise, every possible sort of speculation would be permitted to enter philosophical discussion. It may be, of course, that, apart from the propositions of mathematics and logic, certainty is impossible to attain, or that there are other kinds of certainty which the philosopher must not ignore. But then we should proceed from the certain to the uncertain or from the kind of certainty which is familiar to us to those other kinds of certainty which have yet to be uncovered. I should like to suggest that a re-examination of the Cartesian position will reveal a way to break out of the circle of reflection in which it appears to be trapped. In what follows I shall develop this suggestion by examining two of the most influential directions taken by phenomenology. I shall argue that, despite the marked divergence in their thought, the phenomenological investigations of Husserl and Heidegger not only require and complement one another, but also point the way to a third direction in which phenomenology must move if it is to preserve the significance of its initial insights and unify its Cartesian and anti-Cartesian tendencies.

I

The fundamental impetus behind Husserl's thought is the desire to render philosophy scientific. By this he means that philosophical knowledge should be apodictic. He does not mean that philosophy should take any of the special sciences as its model, for each of the sciences rests upon certain presuppositions which are themselves in need of philosophical clarification and justification. Philosophy, then, is left wholly to its own resources in its search for knowledge. Hence it must seek as its starting-point something which we must acknowledge as absolutely given and indubitable. Like Descartes, Husserl maintains that such givenness is available only within consciousness. Unlike Descartes, however, Husserl does not seek to derive from the *cogito* any knowledge of an independently existing world. On the contrary, he holds that the chief barrier to knowledge is our natural belief that the data of consciousness refer to real existing things which transcend consciousness. So long as we persist in our natural inclination to view knowledge as a relation between the data of consciousness and an external world, knowledge is impossible, for we can never check things as they appear to us against things as they are in themselves. Hence if we are to obtain apodictic knowledge, we must exercise the phenomenological *epoché*, we must suspend our belief in the existence of a transcendent world and focus our attention upon the immanent data of consciousness as they are immediately given to us.[1]

Unlike the Cartesian doubt, the *epoché* is never retracted. But there is no need for a retraction, since the suspension of our belief in the existence of the world has no effect upon the content of experience. This is so, Husserl maintains, because

260

consciousness is fundamentally *intentional*; it is always consciousness *of* an object. The world, then, remains as a phenomenon for consciousness. And if we are content to know *that* we are intending the world and to know *what* we are intending, an absolute science of objectivity is possible. Furthermore, our experience is infinitely richer after we have bracketed the thesis of the natural standpoint, for now no single type of phenomenon is taken as enjoying a position of epistemological privilege. Perception, conception, imagination, anticipation and recall all yield data which are of equal interest to the phenomenologist. Thus Husserl claims to have avoided the reductionism of Humean empiricism and Cartesian rationalism while retaining the insights of both positions.

The phenomenal field left by the *epoché* must now be subjected to a detailed analysis and clarification. At a first glance this field appears to be nothing more than a continuous flux of singular data about which it is impossible to *know* anything at all. Yet this is not the case, for in the judgments we make about these data we have already gone beyond them. In order to relate the *cogitationes* to one another, a universal element must also be given in experience. And if our judgments are to be apodictic, this universal element must be given as absolutely as the *cogitationes* themselves. But how is this possible? Doesn't the universal by its very nature transcend consciousness? To be sure, the universal transcends the *real* data of consciousness, but this is a transcendence which takes place within the immanence of consciousness itself. Hence it is not affected by bracketing the transcendence of the existing world. The real data of consciousness, the *cogitationes*, are profiles through which the universal is given. The universal transcends these singular data as the constant source of identity which effects their synthesis. Thus, in contrast to the reality of the *cogitationes*, the universal is given as irreal or ideal.[2]

The task of phenomenology, then, is to clarify every aspect of the phenomenal field, its objects as well as the acts of consciousness which intend those objects, by grasping the essential structures which give it its meaning. To this kind of operation Husserl gives the name '*Wesenschau*' or 'essential insight.' Essential insight is not to be confused with inductive generalization. When I generalize, I look at a number of individuals, note that they have something in common and isolate that common feature as 'what all such individuals have in common.' To obtain an essence, however, I need not inspect a number of individuals at all. I need inspect only a single individual, vary it in my imagination to determine what remains identical throughout these variations, and *abstract* its essential nature. The mind has the power to abstract as well as the power to generalize, and it is Husserl's contention that abstraction is logically prior to generalization, since the generalization cannot proceed until we have isolated the essence which is to be looked for as the general character of a class of individuals. And this is why phenomenology, as the science of essences, is the foundation of all the other sciences.

The introduction of *Wesenschau* leads us to ask just how we know when we have grasped an essence. The answer lies in Husserl's theory of evidence, which I can briefly indicate here. We must distinguish, he points out, between those intentional acts which are filled and those which are only partially filled or empty. For example, in lecturing on Descartes, I may refer to the triangle as an example of what Descartes means by a clear and distinct idea. To be certain of the truth of what I am saying, I would have to intuit the essential nature of Euclidean space and *see* that a triangle constructed in that space must necessarily have the characteristics which I attribute to it. In so far as I have not brought the triangle to intuitive clarity, I intend it only casually; my intention is not filled, and the evidence for my claims is almost totally lacking. Husserl's theory of evidence, then, is open to all those objections which may be brought against every form of epistemological intuitionism. But to these objections Husserl would reply that if there is any apodictic knowledge at all, it must be first knowledge. Such knowledge cannot be deduced from further premises but must be given unconditionally in and through itself. Phenomenology, then, is a quest for a being which is given to consciousness in such a way that the impossibility of its being given in any other way is grasped intuitively or seen. A clarification of the essential nature of any phenomenon is therefore an assurance that my knowledge of that phenomenon is universally valid; that it holds for all possible subjects.

But I cannot make this claim so long as the experience with

which I am dealing remains the experience of *my* individual worldly subject. What is needed is a further purification of experience—a reduction which brackets the worldliness of the individual subject and reveals the structure of subjectivity in general. Only when the essential structures of the phenomenal field left by the *epoché* are seen as the intentional objects of an equally essential subject can I claim universal validity for the results of my investigations. As a result of this reduction of the subject pole of experience,

I am no longer interested in my own existence. I am interested in the pure intentional life, wherein my psychically real experiences have occurred. . . . We have to recognize that relativity to consciousness is not only an actual quality of our world, but, from eidetic necessity, the quality of every conceivable world. We may, in free fancy, vary our actual world, and transmute it to any other which we can imagine, but we are obliged with the world to vary ourselves also, and ourselves we cannot vary except within the limits prescribed to us by the nature of subjectivity. Change worlds as we may, each must ever be a world such as we could experience, prove upon the evidence of our theories and inhabit with our practice. The transcendental problem is eidetic. My psychological experiences, perceptions, imaginations and the like remain in form and content what they were, but I see them as 'structures' now, for I am face to face at last with the ultimate structure of consciousness.[3]

Furthermore, knowing is not a passive inspection of essences. It is an *act* of the transcendental subject which, through repeated application of the phenomenological reduction, 'constitutes' the essential structure of the world. But if constitution is not entirely arbitrary, it must obey certain laws. And if we are to discover these laws and clarify the meaning of constitution, it will be necessary to grasp the essential structure of the transcendental subject. A pure subject cannot be grasped as an object. But because of the intentional correlation between subject and object, we may take the constitution of the object as a clue to the constitutive activities of the subject. Unfortunately, Husserl never tells us what laws govern these activities. Instead, he describes the activities of a concrete subjectivity in the hope that his reader will recognize the essential laws which they obey. In my consciousness of my own world I am given to myself as a subject in the evidence of my own experience.

In so far as my experience develops according to certain laws, I am aware of the laws which govern the activities of the transcendental subject which is the essence of my own subjectivity. What, then, does my experience reveal to me?

Experience, we have seen, is a continual flux. But the phenomenological reduction reveals throughout this flux the presence of certain essential structures which give an ideal unity and meaning to the stream of *cogitationes*. The constitution of these structures is, according to Husserl, a function of consciousness acting according to the laws of reason.[4] The fundamental law of reason is the principle of self-identity. Only the self-identical can be an object of knowledge. The rational consciousness, then, constitutes objectivity by reducing the flow of experience to what is given in it absolutely according to the principle of self-identity. But because an object is always more than what is *really* present to consciousness at any one moment, its essential structure must be constituted in a successive actualization of its aspects. Consciousness itself is a continual temporal synthesis. It is always one and the same consciousness intending self-identical essences throughout the flow of time. It is essential, then, that the consciousness which effects the synthesis of identity should conform to the fundamental form of all synthesis, the inner consciousness of time.[5]

But to recognize the role played by time in the constitution of objectivity is, at the same time, to recognize that knowledge is a growth. Not every cognition is possible to every subject at every time. For each concrete subject there must be a genesis of cognition. And because consciousness is *essentially* temporal, there must also be a growth of the transcendental subject, the ultimate source of universal validity. Thus with the introduction of time and genesis as themes requiring phenomenological clarification, Husserl is led to introduce the themes of history and the *Lebenswelt*—a world of lived experience constituted by a plurality of concrete subjects.[6] The *Lebenswelt* is referred to by Husserl only in his later writings, but it is clearly in the background of his thought from its inception. It is implied in the concept of universal validity, for a universally valid cognition is one which holds for *all possible subjects*. Husserl's theory of evidence also entails the meaning of an intersubjective world. It is only through my communication

264

with others that I come to borrow words which refer to objects which I have not yet constituted. Hence it is only through my communication with others that the possibility of error can be understood.

The introduction of the *Lebenswelt* as the ultimate theme of phenomenological analysis does not, however, indicate a return to the natural standpoint. The *Lebenswelt* is the correlate of the naive consciousness of the natural standpoint. It is simply there or taken for granted. Philosophical reflection upon this world reveals that though it is pre-reflexive, it is none the less constituted in its being; i.e., the being it has in consciousness has its source in consciousness itself. And when this world is rendered thematic in a rational investigation, it is *reconstituted* according to the laws of reason as an essential world. Thus Husserl remains faithful to his original program. It does not matter whether such a world exists. What matters is that the phenomenologist knows what this world is and that his knowledge of it is scientific because it grasps its essential structure. Phenomenology, then, remains an attempt to guarantee the objectivity of the world by constituting it in a consciousness whose necessity is absolute.

Yet despite the sweeping promises Husserl makes for phenomenology, even the most sympathetic reader will wonder whether he has achieved all that he set out to accomplish. An insight into the essence of subjectivity may tell us how any concrete subject will grow and acquire knowledge, but it tells us nothing about this or that individual *qua* individual. As Husserl himself put it early in his career: 'For phenomenology, the singular is eternally the *apeiron*. Phenomenology can recognize with objective validity only essences and essential relations, and thereby it can accomplish . . . whatever is necessary for a correct understanding of all empirical cognition and of all cognition whatsoever.'[7] But can it really accomplish this if its ultimate task is to grasp the structure of a life-world which is constituted by a plurality of concrete subjects, i.e., individual persons? Furthermore, when we turn to the constitution of another subject, even greater difficulties appear. Husserl cannot appeal to any causal explanation of the other, for such an explanation would transcend the immediate data of consciousness. Hence there must be some way in which another subject

265

is directly given to me, some third form of constitution which is neither that of a subject nor of an object but of a subject objectively presented. Despite its extreme complexity, Husserl's description of the constitution of the other comes down to this:[8] I know my body among all other bodies as a living body. The first determination of another subject is merely that it is a body. Through its behavior this other body is perceived as similar to my own body. And, though this body is a modification of my own perceptual field, it is also given to me as a determinate ego having its own correlative world. It is possible to constitute the meaning of the other and his world because I can comprehend the other as a subject having the experiences I would have if I were *there*. But such an explanation is really no explanation at all, for it has circumvented the very phenomenon it was to explain. The other is not merely an extension of my ego; it is the essentially *alien*. No matter how I try to grasp the meaning of another subject by imagining the experiences I would have if I were *there*, the *there* always disappears in becoming a *here* within my imagination. As my knowledge of the other's situation grows, his situation inevitably merges with mine, and I see him as nothing more than an extension of my own undeveloped potentialities.

Thus the concrete subject remains an *aporia* in the quest for universality and the meaning of another ego is incomprehensible from within the solipsism of the *cogito*. Yet a plurality of concrete subjects is essential if we are to make any sense out of the *Lebenswelt* and out of Husserl's conception of universal validity and his theory of evidence. Had Husserl made the *other* essential to self-constitution in the manner of Hegel, or had he introduced another level of experience to account for the person in the manner of Kant, he might have escaped from this dilemma. But neither of these solutions was open to him because of certain presuppositions which he accepted uncritically. The first of these is the assumption that the theoretical attitude of the rational consciousness is the fundamental form of all cognition. The second is Husserl's tacit acceptance of the *cogito* as the first and only certitude. Phenomenology, then, turns out to be infested with the paradox that it is grounded in a participation in the *Lebenswelt* for which it cannot account. Thus the suspicion arises that the theoretical standpoint of transcen-

266

dental phenomenology is not fundamental and that there must be some other mode of access to the phenomena in question.

II

Heidegger begins his phenomenological investigations with the persistent fact that consciousness is always personal and that all human activity, theoretical or practical, takes place within the *Lebenswelt*. Thus he names man *Dasein* (being-there) to indicate that the individual always finds himself in a specific environmental world which determines the horizon of his activities. The first thing we notice about *Dasein* is that it is not a being whose nature is fixed in the manner in which we ordinarily think of things like chairs or stones as having a fixed nature. It is a being whose being is always in question; a potentiality to be, always in excess of what it has been, always projecting itself toward what it is going to become. Its essence lies not in the static qualities which characterize things but in the dynamic character of its *existence*.[9] Hence we can hope to understand *Dasein* only if we can bring to light its existential structure—those characteristics which reveal how man is placed in the world and what possible ways of existing are open to him. In order to elucidate these basic existentials of *Dasein*, Heidegger proposes to examine man as he is ordinarily given to himself, in the *Lebenswelt*, as a being-in-the-world.

The 'world' becomes accessible as an object of phenomenological investigation when we take as our clue man's primary relation to things within the world. This relation, as Husserl himself points out, is practical, not theoretical. In daily life we do not normally view things as merely present-at-hand (*vorhanden*), as objects of a theoretical inquiry which seeks to comprehend things 'objectively' as they are in themselves. We grasp them as implements or pieces of equipment which are ready-to-hand (*zuhanden*) for the accomplishment of certain tasks within our environment. In contrast to the objects of theoretical inquiry, an implement *qua* implement has no meaning or being as it is 'in itself.' It *is what it is* only within the context of meaning constituted by the project of work in which it is involved.

For the most part, however, this context of meaning goes

unnoticed. It is only when something interrupts my work, when an implement is missing or turns out to be unsuited for its task, that this context is rendered explicit.[10] Suddenly my whole project, the complex 'world' of my activities, is thematized by this interruption; and I am brought to realize that the implements with which I have been working derive their meaning and being as implements through the uses to which I have put them. Furthermore, the world revealed by man's practical interests is not limited to this or that particular project. The product of our work is itself made of something and in this way 'nature' is revealed to us—not as the object of the natural sciences but as something which is useful for something and to someone. A sheltered railroad platform, for example, refers to the individuals who built it and to those who will use it. It refers to the public world constituted by the communal activities of *Dasein*.

The world, then, cannot be understood as the totality of its objects or as the container in which these objects are encountered. It is rather the horizon or context within which objects first appear *as* something. The world does not have the being of an object but the being of a project, which is to say that the world is essentially human. Even the mathematical interpretation of the world is based upon the way in which nature is projected— upon letting nature become involved in a specific interpretation of its being. But the world is not primarily the world projected by the sciences. It is the life-world constituted by our practical interests. Thus distance in the *Lebenswelt* is not measured mathematically but by the projects in which we are engaged. What is measurably closest to me is often so remote in my environment that it escapes my notice altogether. The telephone which I use to call a friend and the street along which I walk are *experienced* as more remote than the person to whom I am speaking or the acquaintance I see in the distance. The world, then, is not to be found among the objects of our concern, for it is presupposed by them. As the project of *Dasein* it is already *there*. It is in fact the 'there' of *Dasein* which makes the disclosure of things within the world possible. And it is through a persistent analysis of the 'there' that Heidegger hopes to uncover the existentials which are the object of his inquiry.[11]

The first of these existentials is disclosed in the fact that I have been *thrown* into a world not of my own making, given over to a situation of which I am not the originating cause but which nevertheless constitutes my past. The way in which *Dasein* is placed in its situation is revealed to it, according to Heidegger, in a vague but telling manner in certain moods. The good and bad moods of which we speak in such a familiar fashion are not merely passing affections. They color our life in the world and constitute an implicit recognition of our *facticity*, of the fact that we are *already-in-a-world* toward which we comport ourselves in a particular manner. Facticity in turn involves a second existential, for every mood is also a kind of tacit understanding. Man is fundamentally a being who exists for certain purposes which he represents to himself. This disclosure of goals for the sake of which the individual exists is, for Heidegger, understanding in its most fundamental sense. It is through its *projects* that *Dasein* first discloses the meaning of the world and of things within the world. The very act of posing a question is such a disclosure, for to question is to sketch in advance the context of meaning in which a particular inquiry will move. *Entwurf* or project, then, is the counter-phenomenon to thrownness or facticity. It testifies to the existentiality of *Dasein* as a being which transcends its facticity in thrusting itself into its possibilities. Thus *Dasein* is never completely determined by the situation in which it finds itself. In so far as it transcends its situation it is free to decide *how* it will develop the potential ways of existing which are provided to it by its situation. But man's facticity is also the source of his alienation from himself, and this reveals the third existential which enters into the constitution of *Dasein*.

The world of daily life requires the individual to fill the position to which he has been assigned. In so far as I am already in the world, the structure of that world is imposed upon me even before I begin to act. It is not you or I or we but the *they* (*das Man*) who determines the goals of my activities. My every action is directed by an anonymous public which, for all of its pronouncements, never reveals itself as anyone or any group in particular. And my inability to assign any personal dimension to this speaker reveals the world of daily life in all of its superficiality and generality. The they is really no one in particular, and under its dictatorship *Dasein* finds it impossible to constitute itself as an individual. The they seduces me away from myself and from the responsibility which my freedom entails by inviting me to obey its orders in return for security, tranquillity and a transfer of my responsibility to it. The more I focus my attention upon things within the world, the more I lose sight of the transcendence of the world, of the fact that the appearance of these things is possible only where there is a free, transcendent *Dasein.* I come to understand myself as a thing and the world as the sum total of its contents. Thus I forfeit an authentic understanding of myself. This inauthenticity of forfeiture (*Verfallensein*) is the third existential of *Dasein.*

These three existentials, facticity, existentiality, and forfeiture, form a single structural whole which Heidegger names 'Care.' Care expresses the intentionality of human existence, and each of its aspects is an existential re-working of certain themes already present in the phenomenology of Husserl. Facticity repeats the theme of the *Lebenswelt*; existentiality echoes the transcendence of consciousness; and forfeiture, as we shall see, is a reinterpretation of the theoretical standpoint of the rational consciousness. But before we go on to this last point, two further phenomena require analysis. Thus far, Heidegger's discussion has remained on the level of generality. But if the *Lebenswelt* is a world of concrete individuals, he must account for the meaning of the person. And this problem coincides with a second task, the constitution of the meaning of authenticity, for the individuality of the person and authenticity (*Eigentlichkeit*) are one and the same phenomenon.

To constitute myself as an individual, I must in some sense grasp the totality of what I am and see that totality as uniquely mine. What is needed, then, is an existential possibility which reveals to me the whole of my existential structure and which is open *only to me.* There is, according to Heidegger, only one such possibility, and that is my death. Death is my ownmost (*eigenst*) possibility because, though another may die in my place, another can never die my death. And since death is also the possibility of my no-longer-being-in-the-world, a comprehension of death should also lead to an authentic understanding of the existential structure of *Dasein.*[12]

In the first place, an understanding of death as my most personal possibility reveals my facticity, for death is not a possibility which is open to me only under certain circumstances or at a particular time of life. On the contrary, I am already thrown into that possibility as soon as I begin to exist. Thrownness into death reveals itself in the mood of dread. Unlike fear, dread has no particular object. Wherever I look I cannot locate its cause or source. It invades me from nowhere, and, wherever I turn, I draw it along with me. In revealing my thrownness into death, dread brings me face to face with my own nothingness as a possibility to be actualized. Furthermore, thrownness into death also reveals that death is not merely something that will happen *to* me at some future point of time. It is something with which I must continually come to grips. In confronting its own nothingness, *Dasein* realizes that it is literally *no thing*, that it transcends the totality of things within the world and is free to determine how it will relate itself to them. Thus my existential projection of death reveals for the first time my situation in the world as uniquely and personally mine. Within the horizon of my no-longer-being-in-the-world I clearly delineate the possibilities which are open to me. Confronted by my freedom, I can no longer interpret myself out of the world. I recognize that I must freely take over my situation in the world, appropriate my possibilities, and bear the responsibility for my decisions.

For the most part, however, *Dasein* flees death in a constant preoccupation with its affairs. In the public world death is seen as a brute fact. 'We all die' replaces the personal relation of *Dasein* to its ownmost possibility. This 'we' is really no one and exists nowhere and for this very reason it can take over my responsibility over and against my death. Yet an unthematic awareness of this possibility makes itself felt in our constant and desperate hurry to fill up time, to keep amused, to stay any return of the self to itself. And this is why, as Schopenhauer observed, the sight of a corpse makes us suddenly so serious.

Despite his fundamental existentiality, man continues to interpret himself as a particular thing within the world. The philosophical expression of this interpretation is that 'reality' is substantial. A substance is characterized by an essential nature which remains identical throughout whatever accidental changes may take place in it. Furthermore, since one substance is distinguished from another by a difference of attributes, and since attributes express the essential nature of a substance, it is difficult to see how two substances can have anything in common. It was under the guidance of these presuppositions that Descartes came to identify the world as extended substance and the knower as thinking substance. The knower and the known, then, are understood as having the *same kind of being*. The transcendence of the world and the existential nature of man are wholly passed over, and a worldless subject is all that remains. How, then, is this worldless subject related to the world understood as *res extensa*? Descartes cannot tell us, for in identifying man and the world as two different substances, he has eliminated any possibility of establishing a relation between them.

Why, then, has this attitude become the 'natural' way in which men think? In equating knowledge with universal validity, the theoretical attitude replaces the individual by the standard observer beloved of the sciences or by the transcendental subjectivity of Husserl. The impersonality of this attitude unmasks it as an epistemological expression of the 'they.' It is grounded in man's alienation from himself, in his inclination to conceal his transcendence, his freedom and his responsibility. Yet despite its claim to being the fundamental form of all cognition, the theoretical attitude reveals its groundlessness by producing insoluble problems like the existence of the external world or by acknowledging its participation in a world of concrete individuals for which it cannot account. What is needed, then, is not a solution to the problem of the external world; not a constitution of an intersubjective *Lebenswelt* from the point of view of the transcendental ego, but the realization that *Dasein* is already-in-a-world, and that the transcendence of *Dasein* *toward* things within the world is the very condition of there being a world at all.

Like Husserl, Heidegger is critical of the traditional conception of truth as *adaequatio intellectus et rei*.[13] In this view only judgments or propositions can be true or false. Within any judgment we must distinguish three components: (i) the real, psychical process which may or may not be present; (ii) the ideal content of the judgment which we call true when it stands in agreement with (iii) the real thing itself. Thus the assertion, 'the picture is hanging askew,' gets demonstrated as true when the individual who makes the assertion turns toward the wall and perceives the picture hanging askew. But what actually gets demonstrated or confirmed in such a demonstration? Let us

272

suppose that an individual makes this assertion without actually perceiving the picture but merely by representing it to himself. To what is he then related?

To 'representations', shall we say? Certainly not, if "representation" is here supposed to signify representing, as a psychical process. Nor is he related to "representations" in the sense of what is thus 'represented,' if what we have in mind here is a 'picture' of that Real Thing which is on the wall. The asserting which 'merely represents' is related rather . . . to the Real picture on the wall. . . . What comes up for confirmation is that this entity is pointed out by the Being in which the assertion is made—which is Being towards what is put forward in the assertion; thus what is to be confirmed is *that* such Being *uncovers* the entity towards which it is.What gets demonstrated is the Being-uncovering of the assertion. . . . Representations do not get compared, either among themselves or in *relation* to the Real Thing. What is to be demonstrated is not an agreement of knowing with its object, still less of the psychical with the physical; but neither is it an agreement between 'contents of consciousness' among themselves. What is to be demonstrated is solely the Being-uncovered [*Entdeckt-sein*] of the entity itself—*that entity* in the "how" of its uncoveredness. . . . The confirmation is accomplished on the basis of the entity's showing itself. This is possible only in such a way that the knowing which asserts and which gets confirmed is, in its ontological meaning, itself a *Being towards* Real entities, and a Being that *uncovers.*[14]

Being-true as Being-uncovering is possible only where there is *Dasein*. Only because *Dasein* is no thing is it free to 'constitute' the meaning of things within the world by providing contexts in which those things are disclosed as something. Taken 'in themselves,' things are not related to one another. Their relation presupposes a process of relating; it presupposes man, not as the thing or subject which *does* the relating, but as this very relating itself. Thus the primordial phenomenon of the truth is not agreement, but the Being-uncovering of *Dasein*. In a secondary sense it is the Being-uncovered of things within the world. This uncoveredness is grounded in the disclosedness of the world. But the world is the 'there' of *Dasein*. Hence in so far as *Dasein* is its disclosedness and discloses, it is already in the truth.

But *Dasein* is equally in error; for truth is commonly misunderstood as agreement and its locus is taken to be judgment. When truth becomes articulated in language, it becomes a body of knowledge which is written down and passed on. The expression of what is uncovered becomes something ready-to-hand which can be taken up and spoken again. For the most part what is known is not grasped through my own uncovering, but through the hearsay of what has been said by others. Truth, then, becomes understood as a relation between two things, subject and object, and its existential character as disclosure is concealed. Thus error, in its primordial sense, is grounded in forfeiture, in man's fascination with the babble of the 'they' and in his tendency to interpret himself as a thing within the world. Through constant repetition, words, which originally articulate what is disclosed, lose their meaning. The contexts in which disclosure takes place come to be understood as categories of a subject. As categories, they lose their power to disclose because they are merely 'subjective.' They can disclose what is only if the subject can make contact with its object. But this point of contact remains a mystery. To understand the primordial meaning of the truth, I must understand myself authentically. Authentic understanding, in turn, means to grasp my transcendence and my freedom, to constitute myself as an individual by grasping my own nothingness. Thus for Heidegger authentic knowing remains the task of the exception—of the poet or thinker who appropriates as his own the contexts and language of the situation into which he has been thrown to use them in a fresh and revealing manner and thus restore their original power of disclosure.

Hence Heidegger, like Kierkegaard and Nietzsche before him, insists that all knowing is, in its deepest sense, existential and personal. Yet unlike his predecessors, he is not content merely to point this out. He wants to go further and develop a *theory* of human existence. But is such a theory possible? We 'know' that *Dasein*'s essence lies in its existence. But what does 'know' mean here? Our knowledge cannot be obtained from the detached point of view of the theoretical attitude. To comprehend Heidegger's thought, we must engage ourselves in an existential project of understanding. *Sein und Zeit* can be fully understood only by a reader who already subscribes to the point of view in question or who is converted to that point of view in the course of his reading. Only then does reading about philosophical thinking become transformed into thinking philosophically.

One might reply, however, that this demand for conversion is implicit in all philosophizing. To comprehend a philosopher's thoughts, we must learn to think them for ourselves. Nor is the appeal to the special reader unique, for philosophy has often been considered as the province of an élite. What is unique in a systematic thinker like Heidegger is his insistence upon the personal nature of knowledge. For Plato or Kant or Husserl the road to truth, though it may be difficult to travel, is open to all who have sufficient intellectual provisions to make the journey. It is not unique and personal but universally intelligible to those who can reason. But for Heidegger it is always *my* road, for it is my project which makes truth as disclosure possible.

Thus Heidegger's conception of phenomenology leads inevitably to a form of philosophical autobiography. My philosophy, it appears, is nothing more than an expression of my understanding of myself and my world. If I think authentically, I make explicit the existential nature of my situation and the possible ways of existing which are open to me. But authentic existence, coming face to face with my own death, also leads to a form of existential solipsism.[15] Hence the direct communication to another of the insights gained through the clarification of my situation seems to be impossible. There is no doubt that much can be learned from such solitary autobiographies, where the task of the thinker is to communicate with his reader not through the 'objective' content of what he has to say, but indirectly, by awakening him to his own existential condition. But this is not how Heidegger conceives his task. *Sein und Zeit* claims to have uncovered how *Dasein* must exist if there is to be anything like *Dasein* at all. Thus it claims the very universality and essentiality which it puts in question. And yet it is difficult to see how Heidegger can justify this claim. In making knowledge personal in the extreme, he has broken up the unity of thought into a plurality of individual *aperçus*. But if there is no independent standard against which we can measure the claims of these different points of view, Heidegger's own pronouncements are, from the point of view of his reader, mere hearsay. Thus phenomenology has come full circle in elucidating its problems. Whereas Husserl was confronted by the paradox of accounting for the individual from the point of view of the universal, for Heidegger the difficulty is reversed. Though he provides us with penetrating insights into the nature of human existence and the *Lebenswelt,* Heidegger has not eliminated the paradox which infests phenomenology. He has merely brought to light another of its dimensions.

III

Universal validity, according to Husserl, is attainable only if we freely submit our experience of the world to the phenomenological reduction. The reduction leaves us with a field of transcendental experience and reveals the Cartesian subject as the transcendental ego, which effects the synthesis of experience by constituting identities throughout the flux of time. But the introduction of time leads to the realization that knowledge is a growth, that not everything is possible to every subject at every time. It leads inevitably to the conclusion that transcendental experience has its ground in a life-world constituted by a plurality of concrete subjects. Heidegger tells us that the concrete individual does not have the being of a subject but that of a project. To know is to permit things to come into relation with one another. Knowledge is possible only because the individual is not a thing but a process of relating things to one another. Thus the theoretical attitude, with its claim of universal validity, is only one possible project of man. Nor is it knowing in its most fundamental sense, for it presupposes the engaged knowing which is a doing and which discloses things as useful for something within the contexts constituted by our projects. These projects authentically viewed are, however, always mine. Hence the impersonal stance of the theoretical attitude is grounded in a knowing which is both existential and personal. This existential attitude, however, must raise itself to the level of universality if it is to take on any meaning or content. Hence each of these attitudes tacitly presupposes the other; each requires the other for its own clarification. A concrete person, then, is neither the dark brooding core of Kierkegaardian existence nor the universal consciousness of Husserl. He is, rather, the continuous interplay of these two factors. Thus man is fundamentally ambiguous, a dynamic unity of two factors which, taken in themselves, appear to exclude one another. How, then, are we to clarify this unity?

276

In another context, E. H. Gombrich has observed that, strictly speaking, it is impossible to *see* an ambiguity.[16] When we are confronted by a figure which can be read in two different ways, like the duck-rabbit figure which Wittgenstein brought to philosophical prominence, we cannot see both shapes at once. If we read the figure initially as a duck, that reading holds us in its spell and we are for the moment powerless to see the figure in any other way. Yet when someone traces out the outline of a rabbit, the duck may be replaced by a second reading which now holds our attention. We come to recognize the ambiguity of the figure in question not by inspecting these two shapes alongside one another, but by continually switching from one reading to the other. I should like to suggest that we become aware of the relation between the theoretical and existential attitudes in a similar manner. So long as I am in one of these attitudes, the ambiguity of my position is concealed. Like the duck and the rabbit, neither attitude can be reduced to the other. Yet they are both expressions of the person under investigation. Unlike the duck and the rabbit, however, these two attitudes not only differ; they appear to cancel each other out. Hence I cannot grasp their relation merely by switching from one to the other. I must instead think through these attitudes to the point at which each reveals its dependence upon the other. This kind of thinking is not thinking about thinking—it is *engaged* thought and it is *dialectical*. Hence we might expect Hegel to provide us with an insight into the phenomenon in question.

It is a basic tenet of Hegel's philosophy that a universal taken by itself is an empty abstraction. The constitution of a state, for example, is a mere scrap of paper with no efficacy whatever until it is accepted by individual citizens as the law which regulates their concrete activities. The individual, on the other hand, is meaningless apart from a content which can be shared by others. Taken apart from my situation in the world I am, as Heidegger points out, really nothing at all. Thus the universal becomes actual through its appropriation by the individual, while the individual becomes someone in and through his appropriation of the universal. But though Hegel has a good deal to tell us about this dialectic, it would be contrary to the whole program of phenomenology to attempt to return to the dialectic of Absolute Spirit. Phenomenology has been forced to acknowledge that the theoretical and existential attitudes, like the universality and individuality with which they are concerned, presuppose one another. Hegel, though he was keenly aware of the dialectical interplay of these two terms, remained on the side of the universal. Essence or the universal is what is cancelled yet preserved in the march of the dialectic—a march in which the individual is ultimately swallowed up. This is why Hegel tells us that time itself must be overcome, that history may go on but that it will have nothing more to teach us. For history is but the temporal image of an eternal process and must be superseded by the eternal knowledge of philosophy. It is also why the free decisions of the individual are ultimately meaningless unless they coincide with the development of Absolute Spirit—a development which is free in the sense that it is not determined by any external factors but follows its own *inner necessity*. Thus if phenomenology is to avoid the reductionism of the Hegelian dialectic, it must seek the clues to the dialectical interplay of the theoretical and the existential elsewhere—in the nature of time, which is for both Husserl and Heidegger the fundamental form of experience, and in human freedom which underlies the two directions of phenomenology in question.

For Husserl time prescribes a necessary order to experience which makes knowledge possible. As in the case of space, Heidegger insists that the clue to the primordial nature of time lies in the *Lebenswelt*. In daily life time is experienced primarily as a 'time-for-something;' it manifests the basic intentionality of *Dasein*. As a being already in the world which is in advance of itself, *Dasein* is an expression of the three moments of time. And because temporality is a continual synthesis of these three moments, it is the source of the existential unity of man. Temporality, then, is the general form of projection itself. Hence the way in which 'time for' is interpreted depends upon the way in which the individual understands his projects. Because *Dasein* is always in advance of itself, authentic understanding sees the synthesis of time as determined out of the future, through the past and into the present. Seen in this light, *Dasein* is the process of temporalization which produces that context of relations which we call the 'world.'[17] Temporality, then, is the fundamental form of all relating which makes knowing in its primordial sense possible.

278

For the most part, however, the individual is taken over by his projects. He does not appropriate his projects and initiate action but waits until the proper conditions for action arrive. Consequently, his understanding of time takes on the character of *waiting for* something to happen *to* him. Thus time, like the world itself, comes to be misinterpreted as a thing; as a continuum of passing moments in which the individual finds himself. The past is taken as a 'now' which is no longer present, and the future as a 'now' which has yet to become present. The three moments of time are levelled to a passing succession of 'nows' in which the individual surrenders all initiative and responsibility.[18]

Yet despite the central role they assign to time, neither Husserl nor Heidegger offers us an adequate interpretation of temporality. Both agree that time originates a context of meaning. For Husserl there is only one such context, for there is one essential order of time which is the same for all possible subjects. According to Heidegger, however, temporality is in its deepest sense personal. Time originates out of those projects which are, authentically viewed, always *mine*. Trapped as they are, each in a different but fundamental direction of phenomenology, they have been unable to grasp these two directions as a unity. Hence they failed to see that time itself provides a clue to the dialectical union of the existential and the theoretical attitudes. Time is, first of all, spontaneous in origin, since the origin of time is time itself. Yet time temporalizes itself according to certain forms. I should like to suggest that the uniqueness of the individual lies in his spontaneous origination of time, while his facticity expresses the different forms which time takes. This dialectic of spontaneity and form becomes phenomenologically accessible when it is understood as the dialectic of freedom.

For Husserl it is only because I am free to alter my attitude toward the world that I can perform the phenomenological reduction. And for Heidegger it is because *Dasein* transcends its involvement with things within the world that it can project itself in a specific manner and thereby constitute a context of meaning within which those things become related to one another. But freedom is never spontaneity alone; it is always freedom to do or to decide *something*. Thus it is always checked by the field of possibilities available to the individual within his situation. Like the posing of a question, a choice of attitude or project gives form and direction to my actions. In choosing to act I choose as well the line of action I must follow if I am to succeed. Furthermore, in so far as they accomplish anything positive, my decisions must be relatively consistent with one another. They are made within a relatively stable horizon of habits and dispositions which has been constituted through my repeated acts of spontaneous origination. Thus freedom manifests the same dialectic of form and spontaneity as time. And because time plays such a central role in the two directions of phenomenology under discussion, we may hazard the guess that the interplay of the theoretical and existential attitudes is an expression of the dialectic of freedom itself—that the dialectic of freedom is the clue to the structure of experience which phenomenology is seeking to disclose.

To lend a degree of plausibility to this suggestion, let us consider briefly one specific kind of experience, the creation and appreciation of works of art. From the point of view of aesthetic experience, each work of art is seen as a unique and self-contained whole. Yet our knowledge of works of art is, to a great extent, general. Each work of art is a child of its time, a product of certain social-historical factors, metaphysical and religious beliefs, as well as the state of contemporary technology. Thus it owes a good deal of *what* it is to factors which are not limited to art but which underlie the whole culture of an age. Furthermore, every work of art is done in a particular style; it is a member of a particular family of artistic forms. Finally, in our critical evaluation of any single work, we make use of principles which claim to be applicable to works of art in general. This general knowledge is not merely secondary to our appreciation of works of art, for where we lack an adequate understanding of artistic style or of the functions a work of art was called upon to serve, our appreciation is often inhibited. This is why, for example, Gothic art appeared rude and barbarous to the eyes of the Renaissance and why Egyptian painting and sculpture was seen for so long as a clumsy precursor to the art of Greece. And yet the more we pursue these 'universal' characteristics, the more we are forced to qualify them. And the more qualifications we add, the more universality becomes dispersed among particular works of art. Neither our knowledge of social-historical factors and of the history of style nor our critical principles ever succeeds in grasping that uniqueness which stems from the hand of the artist. This is so, it seems to me, because artistic genius, as it has often been called, is identical with the spontaneous origination of freedom. Genius cannot be explained theoretically because it is free in origin, an expression of that spontaneity without which the universal elements in art would have no actuality. Spontaneity, however, always requires limits, resistance, and a concentration of its powers if it is to accomplish anything. It is not surprising, therefore, that artistic invention often flourishes where it labors under the strictest of conventions, in Romanesque sculpture, for example. Where no external conventions limit spontaneity, the artist finds himself limited by the nature of his materials, or he imposes specific limitations upon himself, as did Kandinsky and Mondrian. Finally, the spontaneity of the artist is limited as soon as he acts. The first stroke of color upon a blank canvas, like the posing of a question, sets up a universe of discourse within which he will seek a solution to his problem.

I do not mean to suggest that from this first act everything follows of its own accord. The interplay of limitation and spontaneity is evident throughout the process of artistic creation. The artist enters into this dialectic the moment he begins to work, for he is already born into particular styles of working which he may appropriate as his own or against which he may react. The movement of this dialectic is present in the act of creation as well, for the artist is alternately engaged as the maker and detached as the spectator of his own work. The work originates from the unique and unrepeatable strokes of his brush, but the direction along which this origin will develop is also conditioned by his critical detachment which makes him a spectator of his own activity. Finally, even critical evaluation is not wholly theoretical. The theoretical attitude always aims at universality; if it proceeded unchecked, it would end by substituting for the work an essay on what the work is. To appreciate a work of art, we must penetrate to the uniqueness which lies in its origin. This origin is present to us in the signs of the artist's activity, in the strokes of his brush or the marks of his chisel. But these signs cannot be grasped conceptually. As soon as we describe them in general terms, the uniqueness which they indicate

281

vanishes. They disclose the creative act only when the spectator becomes, for a moment, the artist; only when he appropriates the work as his own through what, for lack of a better expression, I shall call an 'existential act of assessment.' If I cannot state *what* this act is, I can at least indicate how it takes place by means of an anecdote. A visitor once called upon Matisse and found him surrounded by a number of drawings. Wondering how the painter could choose among them he asked Matisse, 'How do you know the really good ones?' 'How one knows the really good ones?' Matisse replied. 'Well, one feels that in the hand.'[19] Only by undergoing the act of origination can we come to grips with the uniqueness of a work of art. Thus our appreciation of works of art is not wholly theoretical. It involves a purely personal dimension, a free appropriation of the act of creation through which we become aware of an originality which cannot be grasped conceptually.

This double role of the artist as spectator and maker is also present in his relation to the finished work. Once a work of art takes on its final form, it is no longer the exclusive property of the artist. Even those traces of its originality, the brush-strokes, are there for everyone to see. Through its total form it achieves a kind of universal accessibility, for it can now be examined, discussed, appreciated, and appropriated by others. In becoming accessible to all, it belongs of itself to no one; it belongs to an individual only through his free appropriation of it. Thus the artist is no more obligated than the viewer to acknowledge the finished work as his own, and it is not surprising to find that artists have frequently disowned or failed to recognize their work. Yet an artist who disowned *each* successive work, like an individual who disowns all of his actions, would end by producing nothing at all. The fashioning of art, like knowing, like life itself, is a continual growth. Where there is no continuity of form there is no development. Thus through the continual interplay of form and spontaneity the artist comes to fashion his style. Because form and style are principles of intelligibility, the artist can be both maker and spectator of his work. And because his forms and his style make his work accessible to others, their origination establishes a universe of discourse, a tradition of viewing and painting, in which communication with others becomes possible.

Like the artist, the individual is thrown into a situation in which various possibilities are open to him. As the artist experiments with different materials and styles, so the individual experiments with different ways of existing. In this process of experimentation he is both creator and critic. He stands in the same relation to his facticity that the artist maintains with his work. When I stand back from my life, my values, my truths, and my hypotheses, to assess them critically, I take the standpoint of another; I acknowledge them as being universally accessible to all. But my critical evaluation is by no means wholly theoretical, for to know *what* possibilities are open to me is not the same as comprehending what it is to make those possibilities actual. To take a possibility as *real*, I must assess it existentially. Thus when I come to assess the alternative philosophical positions offered me by Plato, Hume, and Kant, it is not enough merely to view these different forms of philosophical thinking from the detached point of an observer. To make a detached thinking about thinking the basis of a decision would be like buying a suit of clothes without first trying it on. Genuine critical thinking must, on the contrary, be engaged thought. It becomes engaged through an existential appropriation of the possibilities in question. To understand these various philosophical attitudes I must become in turn Plato, Hume, and Kant. I must make their thoughts my own by learning to think through them on my own. Hence a purely personal dimension underlies the most abstract thinking. It is this spontaneous act of appropriation upon which the existential attitude lays so much emphasis. Through repeated acts of appropriation I originate a horizon of dispositions and habits which constitute my style of life. But a style of life is no more a purely private affair than a style in art. Its constitution takes place within a cultural horizon which is already public. And in constituting for myself a *form* of life within this horizon, I both constitute myself as an individual and enter as well a universe of discourse which makes communication with others possible. The authentic constitution of the person, then, does not consist in sacrificing form to spontaneity, but in holding fast to the dialectic of freedom in which such constitution is first made possible.

As first philosophy, phenomenology must deal in first principles. But the first principles of Husserl and Heidegger, the transcendental subject and the concrete person, presuppose one another. Thus phenomenology must take as its first principles the dialectical interplay of these two factors or the dialectic of freedom itself. In making freedom its proper object, phenomenology establishes itself as first philosophy in a twofold sense. Every other form of knowledge develops within specific attitudes which man makes toward the world. These attitudes are never put in question, for the questions we ordinarily raise constantly presuppose them. Phenomenology liberates itself from the domination of any one of these attitudes by recognizing the spontaneity of freedom. In recognizing that these attitudes can be *taken* only where there is a taker who is free to appropriate them and thus originate a context of meaning, phenomenology comes to understand that the unity of experience cannot be reduced to any single principle. This unity is grounded only in the continual presence of those forms which originate spontaneously and which render it intelligible. Finally, in dealing with freedom, phenomenology establishes itself as the philosophical method *par excellence*. For as a spontaneous origination of form, freedom is not subject to further explanation. It is an absolutely first principle which can only be *described* in its operations.

In suggesting that the structure of experience is constituted by the dialectic of freedom, I do not mean to imply that this structure is entirely arbitrary. The contexts of meaning in terms of which we seek to understand the world exhibit a remarkable degree of uniformity and consistency. It is one task of phenomenology to investigate these different attitudes which man takes toward the world and to exhibit, if possible, their interconnections. But these attitudes are not entirely the product of man. Just as the artist meets resistance in his materials, just as his materials react with and contribute substantially to his forms, so things often resist our attempts to deal with them from a particular point of view. Thus we find ourselves forced to alter our universe of discourse by the nature of the things themselves. Hence the task of phenomenology is twofold. On the one hand, it seeks to discover the modes of understanding which man brings to experience. On the other, it investigates the resistance of phenomena to the contexts we seek to impose upon them. Finally, I have suggested that the continual origination of

WILLIAM H. BOSSART

form which ties experience together is fundamentally temporal. And if this suggestion is borne out by further analysis, phenomenology will ultimately become the morphology of time—a study of the shapes of time themselves.[20]

REFERENCES

[1] E.g., E. Husserl, *Ideen zu einer reinen Phänomenologie und phänologischen Philosophie, 1*, Haag: M. Nijhoff, sec. 27–32. Eng. trl. by W. R. Boyce Gibson, *Ideas*, London: Allen & Unwin, 1931.
[2] E.g., Husserl, *Die Idee der Phänomenologie*, Haag: M. Nijhoff, 1954, Vorlesung IV.
[3] Husserl, 'Phenomenology,' in *Encyclopedia Britannica*, 14th ed. Cf. also *Cartesian Meditations*, sec. 11.
[4] *Cartesian Meditations*, Meditation III.
[5] *Ibid.*, sec. 18; also *Ideas, 1*, sec. 81.
[6] *Cartesian Meditations*, Meditation V; also *Die Krisis der europäischen Wissenschaften und die transzendentale Phänomenologie*, 2nd ed., Haag: M. Nijhoff, 1962, pt. III A.
[7] Husserl, 'Philosophy as a Rigorous Science,' in *Phenomenology and the Crisis of Philosophy* (trl. Q. Lauer), New York: Harper Torchbook, 1965, p. 116.
[8] For a full discussion of the constitution of the other, see the fifth Cartesian Meditation.
[9] M. Heidegger, *Sein und Zeit*, 7th ed., Tübingen: Niemeyer, 1953, sec. 3, 9. Eng. trl. by J. Macquarrie and E. Robinson, *Being and Time*, New York: Harper, 1962.
[10] E.g., *Ibid.*, sec. 16.
[11] *Ibid.*, Division I, Ch. IV, V.
[12] *Ibid.*, Division II, Ch. I.
[13] *Ibid.*, sec. 44.
[14] *Ibid.*, pp. 217–18 (pp. 260–1, Eng. ed.).
[15] *Ibid.*, p. 188 (p. 233, Eng. ed.).
[16] E. H. Gombrich, *Art and Illusion*, New York: Pantheon Books, 1961, pp. 4 ff.
[17] Heidegger, *op. cit.*, p. 365 (p. 417, Eng. ed.).
[18] *Ibid.*, p. 424 (pp. 476–7, Eng. ed.).
[19] Cited in E. Gilson, *Painting and Reality*, New York: Meridian Books, 1959.
[20] I owe this phrase to George Kubler's *The Shape of Time*, New Haven: Yale, 1965.

11

PHILOSOPHICAL KNOWLEDGE OF THE PERSON

Edward Pols

I

One way of opposing a reductionist view of life in general and of man in particular is to point out realities that such a view is unable to explain in fact, and then to show that in principle these realities must remain outside its scope. If one wishes to go beyond these negative considerations and offer alternative non-reductionist explanations of these realities, one meets with at least two difficulties. In the first place, such explanations do not conform to the dominant ideal of exact scientific explanation. In the second place, they often seem to resemble metaphysical explanations. On either count their credentials are in question.

It seems to me a mistake to attempt to offer, in response to these difficulties, more generously conceived versions of scientific explanation that have been purged of metaphysical overtones, because I believe that all modes of scientific explanation remain methodologically obscure without metaphysics, and that some of them *should* contain a metaphysical element if they are not to distort their subject-matter.

This is especially true when the subject of our study is ourselves, as it is in the present paper. My topic is the Person: its nature, the nature of our knowledge of it, and the interdependence of that knowledge and metaphysical knowledge, or, as I shall also call it, philosophical knowledge. Scientific and quasi-scientific modes of knowing have of course largely pre-empted the word 'knowledge'; if this should prove a stumbling block to any reader, it will not seriously distort my intent if the word 'understanding' is substituted for 'knowledge' throughout the paper.

The interdependence of metaphysical knowledge and knowledge of the Person is a matter of some delicacy. When we address ourselves to the topic of the Person, we are addressing ourselves to the study of man. In that study we seek *to know* (sense 1) an entity, or being, a great part of whose vocation is *to know* (sense 2). It seems reasonable to distinguish the two senses thus, since there will be at least some situations in which they will not be the same. Certainly we apply the verb 'to know' to a variety of activities, to some willingly and to others grudgingly, but it is by no means clear that any or all of them adequately define the sense in which it is man's vocation *to know* (sense 2). And it is still less clear that any or all of them wholly define the reflexive sense in which we seek to know ourselves as knowers (sense 1).

This, it appears to me, is not just a sign of obscurity and confusion. It is also a sign that we are dealing with a normative issue. We are looking for a mode of knowledge that we *ought* to possess, one that we are in fact trying to bring about. But observe that a sense of 'to know' that would fit this requirement would also fit either sense 1 or sense 2. And in both occurrences it would stand as a norm against which we might judge other senses of 'to know.' The issue, therefore, is not only a normative one but a reflexive one as well. But from the beginning metaphysical knowledge has been thought of in just those terms. Consider first the normative issue. Metaphysical knowledge has been directed upon Being: to know *that* would be to possess a norm or standard in terms of which other modes of knowing might be judged. It has, moreover, been directed upon Being as upon a question, for it is not clear what we ought to mean by that word, and it is still less clear whether our cognitive powers are fitted to attain such knowledge. And with this last question the reflexive issue enters, since to settle it we must in some sense come to know the nature and extent of our rational powers. Both issues come to a head if one considers the Person as a being or entity. To gain an appropriate knowledge of ourselves as entities: how can we do that without becoming the entities, the beings, we ought to be?

In Part II I shall give an account of a metaphysical (or philosophical) knowledge (or understanding) of the Person that I hold to be accessible to us. I say 'accessible' because I believe that it is present in all of us in a primordial and partially latent form, and that we must reflect upon it to develop, clarify, and authenticate it. In this phase of the discussion, however, it is important that that supposed entity, the Person, serves *only* as an example of the fundamental sense of 'entity' we are seeking. The point is that this mode of knowledge applies to all entities whatsoever: I shall maintain that it is a knowledge of entities (beings), of Being, and of causation. 'Being' will be taken in a very concrete sense; the capitalization is deliberate and will not be used, for instance, in the expression 'being of an entity.' 'Causation' will be taken in a sense both more concrete and more manifold than is usual in philosophy of science. The connection between the categories 'being' and 'causation'—a connection frequently obscured in modern epistemological discourse—will be emphasized by way of the notion 'ontic power.' The words 'being' and 'entity' are of course ubiquitous and ambiguous. Since we shall be looking for a very fundamental sense of these words, we shall not be concerned either with determining all of the things we shall or shall not apply the terms to or with exhibiting the connections between our fundamental sense and all other senses.

In Part III the claim is made that metaphysical knowledge is a *framework* knowledge, in which such questions as the relation between reason and experience are answered differently from the way they are answered for more specialized activities that take place within the framework, and in which topics such as causation take on a more concrete and complex meaning than they do for these more specialized activities. We may say either that Being is our framework or that any entity we focus upon in this way is our framework.

A first step towards the authentication of this framework knowledge is undertaken in Part IV. Authentication of it is shown to be dependent upon a knowledge of ourselves as persons, more particularly, upon knowledge of those functions of the Person that enter into it. These functions, however, are very complex, for the very philosophic impulse that moves us both to seek metaphysical knowledge and to seek to authenticate it involves the whole affective structure of the Person. Self-knowledge of this sort is, moreover, both congruent with the metaphysical knowledge we seek to authenticate and interdependent with self-integration (at least in the sense that to distrust this mode of self-knowledge is to create hazards for self-integration). Authentication, then, is also both reflexive and normative.

In Part V authentication of this mode of knowledge is held to be inseparable from an ontic expansion of the Person and its rationality. This expansion is held to consist in part of a creative (or radically originative) reflection. A brief sketch of this doctrine ends in a return to the theme of Being and causation.

II

Many of the issues towards which a discussion of the Person might naturally be directed were once discussed in terms of the reality and nature of substance. Is the body a substance? Is the mind a substance distinct from it? Is the soul (or the Self, or the Person) itself a distinct substance or a composite of the other two? The use of the word 'substance' to cope with such problems is unfortunate. Behind the word lies the Greek word '*ousia*,' of which 'substance' is certainly no happy translation. '*Ousia*' can be more exactly rendered as 'being,' or 'entity,' although the overtones are lost unless we also keep it in mind that in searching for what *really* deserves to be called *ousia* we are in fact searching for a very fundamental sense of the word 'being,' or 'entity.' The use of the word 'substance' in the conduct of this search gives rise to many obscurities, chiefly because of its literal sense—'that which stands under.'[1] Some such obscurity certainly lies in the Cartesian notions 'thinking substance' and 'material (or extended) substance,' which have dominated so much of modern inquiry. They tempt us towards the conception of an underlying, permanent, homogeneous stuff: we think of consciousness as something at once diversified by and hidden by particular thoughts, feelings, and volitions; and think of material substance as something at once diversified by and hidden by the sensory qualities in terms of which we know it. If we raise the question whether the Person, or the Self, is a substance, we find ourselves answering it in the same way. In all cases what we are looking for becomes 'an "x" I know not what,' although the logic of our argument makes us think of it as an inane and neutral 'x' that it would be profitless to know even if we could do so.

Once established, the notion of 'an "x" I know not what' is hard to shake. It is hard, indeed, to avoid thinking of the Kantian thing-in-itself in such terms. It seems fair to say that the thing-in-itself stands in Kant's work for the *real* substance that philosophers before him sought, as he believed, so mistakenly to know; while what we grasp by means of the *category* of substance falls short of true reality, which must remain featureless for the theoretic intelligence.

Let us try to disburden ourselves of some of the difficulties of the traditional term, 'a substance,' by using instead the term 'an entity' or the term 'a being.' This has at least the merit of directing our attention towards the plurality and individuality of things, while at the same time preserving, through the connection with the term 'Being,' an openness to the generality and unity we point to by that ancient word. There is the additional merit that we shall not set impossible and stultifying standards for the knowledge we are looking for. Dismissing the image of an underlying something or other whose purity and importance is, as it were, defined by an emptiness of content will permit us to value the concreteness and individuality of things as positive expressions of Being. We shall not demand that if something is to have the status of an entity there must be discoverable behind its (phenomenal) concreteness an inane, neutral, and therefore ineffable core. And we shall not be disappointed that our knowledge of Being is refracted by the particular entities in and through which we know it, for we shall not be supposing that Being is an inane, neutral, and therefore ineffable core of things. The reader needs no reminding that Aristotle in his search for Being *qua* Being had to look for it in the being of the individual entity or *ousia*. No doubt he did not see so clearly as we all hope one day to do, but it is at least plausible that he was looking in the right direction, always providing, of course, that entities are indeed there for us to know, identify, and dwell upon.

Philosophy has of course never quite lost sight of the expressions 'an entity' and 'a being' in its preoccupation with the problem of substance during the epistemological development from Descartes through Kant.[2] The objection that these terms are ubiquitous and unavoidable and hence confusing is not to the point. For if it is confusing to say that the route to Being is through the being of a being—and who would disagree with such a complaint?—it might well be that such a claim is precise, true, and illuminating when taken in a well-wrought context.

That any such context must be a difficult one is true enough, but where has any important simplicity been established except in a context of some difficulty?

The claim then is that we are capable of knowing entities, capable, that is, of judging something to be an entity (of a fundamental sort), of according the status of entity to what might be called from some other point of view an experience or set of experiences, a phenomenon or set of phenomena, an event or set of events. The detailed account of this claim will involve the further claim that our knowledge of the multiple entities of the world is also a knowledge of what philosophers have called Being and a knowledge of a fundamental sense of causation.

Let me begin, however, by mentioning one concession to the Kantian approach that, it seems to me, ought to be made by anyone seeking to claim that we have a knowledge of entities and causation. Although Kant was not right in claiming that we have a synthetic *a priori* knowledge of substance and causation of the kind set forth in the first critique, he was absolutely right in supposing that, if we are to know substances (read 'entities') and causation in any defensible sense of 'know,' then there must be some universality and necessity in our knowledge. This is not, to be sure, to claim that we shall not in the future be surprised by the character of entities we encounter and the forms causation can take; nor is it to claim that in order to know an entity *qua* entity or a cause *qua* cause we must have so plumbed their natures that our understanding of these terms is incorrigible by future philosophic efforts. It is simply to claim that what one understands of an entity *qua* entity or a cause *qua* cause will illuminate to the same degree any entity and any cause whatsoever. Perhaps there is no philosophic knowledge of entities and causes, and the thesis of this paper quite mistaken; but, if there is any such knowledge, there must be some universality and necessity (in at least the modest sense just mentioned) to our knowledge.

It should be noticed first that philosophic knowledge of what it means to be an entity is *sui generis* in the sense that we cannot make it clearer by thinking of it as built up out of more immediate and more familiar elements. Consider some of the things that have been said, and said rightly enough, about entities: an entity remains in some sense permanent throughout the changes it undergoes; an entity is a kind of unity in multiplicity; entities differ in the kind of order they manifest—there are degrees of being and some entities manifest a higher degree than others, and some manifest several degrees within themselves; an entity represents an achievement of a given level of order— it is a mode of activity, and because of this we cannot understand it without introducing the idea of value; an entity is the form a process takes, or the structure that joins a number of disparate elements. These claims may all be true enough, taken with appropriate qualifications, but none of them and no combination of them *defines* an entity for us. It is more accurate to say that we already know what we are looking for and that we recognize these as legitimate aspects of what we are trying to clarify. We are in fact trying to point to something so fundamental that what might at first glance seem more simple or more fundamental shall appear as in fact an attenuation of what we are interested in. Any clarification we can give must presuppose that we can recognize an entity, that in some sense we know what we are looking for.

There is, however, a clarification we can give that is not based upon an attempt to build up the notion 'entity' out of something more fundamental. It is this: that in our recognition of any *complex* entity we are capable of recognizing some, at least, of the subordinate entities that contribute to it. The recognition is double: we recognize the subordinates as entities, and we recognize that they do indeed contribute to the being of the more complex entity. The recognition is double in another sense. The complex entity is not merely a *complex* of entities: it is itself a complex *entity*. Already, then, at least one aspect of the problem of causation is in view, for the contribution of the subordinate entities is in some way causal.

But when we are dealing with a complex entity we do not merely recognize it to be made up in part of a complex of particular subordinate entities. That an animal's body consists in some sense of an assemblage of cells at any given moment; that its history is the history of a developing assemblage of cells; these are true enough. But it is also possible to regard it as a complex of levels as well, and to regard its history in terms of a gradual development of successive levels. We may think of these as levels of operation, levels of function, or levels of order. To

U   293

speak of levels is to introduce immediately an issue that in quite another context might be dealt with in terms of the problem of universals, for a level is something that is by definition general. It is not (to continue with the biological example) just a particular assemblage of particular cells: it is a certain *kind* of assemblage of certain *kinds* of cells. That in a specific instance we identify it as a level means that we could identify it in other instances of its occurrence. This is an important ontological consideration since in recognizing any entity we are at once recognizing something unique (this particular animal, this particular cell, or this particular molecule) and something general (the type or level to which the animal, cell, or molecule belongs).

We shall return to this a little later. For the moment it may be helpful to give an example of the doctrine of levels in terms of the chief subject of this paper, the Person. Any person exhibits a layer of atomic order; molecular order in general; the special type of molecular order that, as in the case of DNA, controls certain processes; cellular order; various layers of structural order as recognized by the anatomist; functional orders in which several structures, and several layers of lower order, play a role, such as digestion and homeostasis; several distinct controlling modes of order, according as we concentrate on later or earlier structures of the central nervous system; the order of consciousness, if this should be in some sense distinguishable from those orders that form its physical basis; an affective or emotional order; a moral order, if this should be distinguishable from the former; and so on. Lastly, the very kind of entity we have under discussion, the Person, is itself a level of order: there are other persons, and persons manifest the level of Personhood.

There are many precedents for this way of regarding a complex entity. The Aristotelian Scale of Nature, in which any given general level is regarded as form in relation to the next lower, and as matter in relation to the next higher, level, and in which at least some individual entities or *ousiai* can be thought of as (formal) complexes of (material) subordinate particular entities, is a case in point. Leibnitz's doctrine of the relation between a dominant monad and its colony would appear to emphasize what I have here called the relation between a complex entity and its subordinate individual entities, but he also gives considerable attention to the hierarchic disposition of *kinds* of entities. Michael Polanyi has restated the doctrine of levels of reality in an original way in our own time and illumined with it many puzzling issues in the philosophy of science.

Seen from this point of view our recognition of complex entities is, on the one hand, a recognition of individual entities and general levels of being that are subordinate, and, on the other, of individual entities and general levels of being that supervene. The recognition of any entity, whether subordinate or supervening, will be *sui generis*. We thus do not mean by 'supervening entity' merely 'complex of entities'; knowledge of the supervening entity is not equivalent to knowledge of a complex of entities. Indeed, it is hard to see why, if we insist on regarding what I here call a supervening entity as simply a complex of entities, there remains any reason for regarding the entities in the complex in their turn as anything more than complexes. The reductive impulse is imperious and will not stop until it reaches ultimate 'particles,' and indeed has no logical reason to stop there.

These claims can be extended from the structure of the entities I claim we know, or understand, to the structure of our knowledge itself. Analysis of our knowledge of any entity may reveal a multitude of discrete sensory experiences, impressions, sense data, qualities, and the like, the terminology depending upon the bent of our epistemology. But no such assemblage can be equivalent to the cognitive awareness in which we recognize that these data signalize the presence of an entity; and none can wholly justify the articulated judgments in which we embody that recognition. That does not mean that these judgments are any less reliable, general, and impersonal. Indeed, I hope to show later that they are, when permeated by a philosophic reflection that is completely congruent with them, self-justifying; and that they then supply, rather than require, epistemological foundations.[3] For the moment my claim is simply that there is no evidence prior to and independent of our capacity for recognizing entities that can be summated to yield the conception 'entity.' Evidence *can* be given for the presence of an entity, and we can then be mistaken in using that evidence, but the *relevance* of the evidence is established by our capacity for recognition.

We turn now to the topic of Being. Entities, or beings, however unique and particular, never present themselves as a bare multiplicity. Any entity whatsoever is rooted in, participates in, shares in, is a partial expression of, Being, which transcends any of its particular manifestations. The claim sounds dogmatic, but I mean it to be taken subject to the same demand for authentication that confronts the supposed knowledge we are now discussing. But it stands, I think, to benefit from any authentication we find for this knowledge. What I am saying is that *to know* an entity in the sense intended is also to know something of the Being it shares in and also *to know* that Being as transcending its instances. And if this were not so, it is hard to see how we could attribute to any entity a transcendence of any particular instant in which we should attend to it. To make a point similar to the one existentialists sometimes make about the transcendence of facticity, it is also hard to see how we could think of any entity as transcending everything that is determinate in its past history. The claim that our knowledge of any entity transcends any assemblage of data is of course inseparable from the present claim.

To make this clearer I must return again to the claim that there is some universality and generality in our knowledge of entities and causation. Our grasp of these things *in so far as it is cognitive* exceeds all the instances we have seen precisely because we see them *as instances* of something that is common and concrete. It is true that any entity is particular and unique: just *this* entity here and now, which when it is gone leaves a metaphysical gap nothing else can fill. But to put it this way is to miss something important. It is also true that I am, as I grasp an entity cognitively, not merely in this unique place and time, do not grasp a merely unique set of data or a merely unique set of particulars, and do not articulate in discourse what I thus attain for my merely unique and particular self. There is in our knowledge of a particular entity something general or common; I do not mean merely in our concept of it, but in the thing as well: it is a being and its being expresses for us something of that which is both concrete and general: *Being* qua *Being*.

I would distinguish this common and concrete character of Being from the *concept* of Being in general: the two should no more be confused than should some individual man and the concept of man in general. One way of arriving at the concept of Being is by stripping away from our concept of some particular kind of entity all that serves to define it and distinguish it from other entities. The concept we end with seems an empty one, and we cannot conceptually distinguish it from Nothing. We shall not here try to untangle the difficulties about the relation between concept and thing, except to say that as the concept 'man' is in some way founded on individual men, so the concept of Being is founded on what I am here trying to get at by speaking of Being as concrete and common. The concept man is 'empty' in respect of the particularity and concreteness of individual men; the concept 'Being' is empty as over against the common concreteness of Being.

There is another way of distinguishing the common concreteness of Being from the concept of Being. Being has often been identified with the creative source of things, which has been called God, the First Cause, the Ground of Being, the Form of the Good, and so on. We may quarrel with the attitudes that give rise to these expressions, but they have the merit from the present point of view of pointing to the power, order, and abundance of Being. Seeing it thus, we see it not as an absence of all characteristics, but as a superabundance that exceeds all particular charcterization. The generality or common character associated with Being then appears in a special light: it is the One from which the Many—the many particular entities of the world—receive that which makes them entities. And if this should be so, it is also that from which a whole class of similar entities receive that which makes them thus similar. If it is legitimate to understand Being in somewhat this sense, then there is no necessary strife between generality and concreteness.

Seen from this point of view, the unity and individuality of the Person (its status as an entity) and the unity and individuality of its components (their status as entities) are alike expressions of a concrete and common Being. There is no warrant then for regarding the Person as merely a complex of entities, for the entities of which it is made up share the same root as the Person and are therefore as problematic as it is. But on the same hypothesis, we cannot regard the Person (or indeed any of its subordinate entities) as *merely* unique and particular, since

297

what confers individuality and particularity upon it is general or common in its concreteness.

The permanence of the self as an entity also becomes something not wholly elucidated by the relative permanence that is the persistence or repetition of a physical pattern through a period of time. Since (on our hypothesis) the Person as an entity derives its being from the source of all permanence, we may as readily claim that the Person *expresses* the permanence of its source in terms of the persistence of a physical pattern and the concomitant unfolding of a physical pattern, as that the permanence of the Person is *caused* by the persistence of a physical pattern. Or, to put it differently, the persistence of a physical pattern is as much explained by the entity we call the Person, as the Person is explained by the persistence of a physical pattern.

But let us not avoid the deepest problem concerning permanence that is raised by the complementary character of the terms 'entity' and 'Being.' If the concrete though general Being we are considering is that in which each entity is rooted as in a source, then it is 'out of' time, in a sense familiar enough from the tradition. We have, therefore, to suppose that an entity that derives its unity, individuality, and (relative) permanence from Being is permanent *through* its process by pivoting for the time of that process about a point that is 'out of' time. This does not commit us to the image of a vapid and static eternity. The point can just as easily be used to restore to our conception of time a vitality it lacks when one thinks of the contents of any present moment as wholly caused by the events in its past. Our point therefore parallels the many efforts made in this century to give process its due. The insistence, in their different ways, of Bergson and of Whitehead on the creative vitality of a present that 'contains' the future in a partially indeterminate way is a case in point. The claim by existentialists that man transcends his own facticity is another case in point, although it should be observed that I am not restricting these claims to the entity we call man. With these qualifications about the expression '"out of" time,' the claim is that any entity whatever, including the Person, does not have its status as entity wholly clarified in terms of a temporal process. Just as an entity, taken as one of the Many, is not merely unique, so an entity, taken as one of the

Many that in their togetherness yield physical process, is not merely a physical process. Each of the Many participates in the One; each changing thing participates in what in some sense does not change.

Let us now return to the earlier claim that a complex entity can be regarded as supervening either upon a complex of subordinate entities or upon a complex of subordinate levels. If we take seriously the suggestion just introduced, that Being is a concrete but general source of order, power, and reality, the subordinate levels must also be regarded as expressions or manifestations of this source. Being is One over against the Many, whether we regard the latter as subordinate individual entities or subordinate levels.

The 'relation' between Being and the component entities and levels of a complex entity is not, however, different from the 'relation' between Being and the supervening entity. The latter and its level, the component entities and their levels, are all of them expressions of Being. Similarly the fusion of particularity with generality (the generality of a common concrete source) is a characteristic of any entity at any level. A supervening entity therefore does not merely bind together and unify in a merely particular collocation the entities that are its subordinates, for what the supervening entity is and does expresses Being. We may just as well say that Being binds together and unifies the subordinates as that the particular supervening entity does. A supervening entity therefore expresses Being throughout the whole of its being, which includes the contribution of its subordinates; alternatively, it expresses Being by means of these subordinates. If we consider a person as expressing himself in a rational order, a moral order, an affective order, the order of consciousness, a supporting nerve-net organization, and a whole series of supporting 'lower' orders, we must think of the total action as that of a unity. But we must now use the word 'unity' in a sense incompatible with an unqualified pluralism. Thus, although the particular entity in question, the person, is one over against the manyness of the subordinate levels and entities in which it expresses itself, its unity *as* a particular entity is *also* an expression of the level of Personhood and an expression of Being itself. The highest entity 'makes use of' the subordinate levels and entities to 'express' itself; and Being 'makes use of' the highest entity and the level of that entity to 'express' itself and, in this way, 'makes use of' the subordinate entities and levels as well. Many of the latter will of course have an existence independent in some sense of the supervening entity, for Being expresses itself in all levels and in all entities. All these points are subtle, depending as they do on changes rung on the word 'being'; a rehearsal of some of them in terms of causation may clarify them.

It will be maintained that an adequate conception of causation is inseparable from an adequate understanding of entities and Being. To take an entity seriously is to take it not just as a *complex* of entities—the effect, that is, of which they are the causes—but as a complex *entity* as well and in that sense, precisely *as* entity, a center of causal efficacy. We can not appropriately think of a (natural) entity (in the fundamental sense we have been pursuing) as wholly caused, precisely because we find it thoroughly appropriate to think of it as exercising causal power or efficacy.

It must be obvious, then, that we are refusing to take seriously two modern dogmas about causation that derive ultimately from Hume. The first is that causation is best discussed in terms of events; the second, that the event or events described as the cause must temporally precede the event or events described as the effect. I shall propose instead two basic and related senses of causation, both of them couched in terms of the category 'entity' rather than the category 'event.' I do not mean to imply that the category 'event' is not for some purposes a valid one. I should even concede that there is a perfectly defensible sense in which we may describe an entity as a complex event. This is especially so if we wish to consider an entity only in terms of spatial configurations and temporal process. Considered thus in terms of what Whitehead called 'extensiveness,' the entity can be analyzed into sub-events, of which the earlier ones will contribute causally to later ones. But, setting aside such a restricted and abstract context, we shall not in what follows be ruling out the simultaneity of cause and effect. It is also important that we shall be taking a radically different attitude to the question how we come to *know* causation than is tolerated by the tradition that goes back to Hume.

It should be observed first that from the earliest times philosophers have recognized a sense of causation alternative to the one expressed in terms of events and temporal priority. It is the sense in which a given level of order, formality, or being has been recognized as *itself* a cause, but in a sense outside the temporal sequence.[4] From the point of view of the temporal sequence, the level of order has been thought of as a goal towards which the sequence moves, and the movement has been conceived of on the analogy of the purposeful movement of an intelligent being towards its goal. On the cosmic plane the causation of Being (as over against finite beings or entities) has been conceived of in a teleological sense as well, and in theological schemes this has been understood in terms of a divine plan. The difficulties in teleological doctrines of causation are manifold, and no doubt such doctrines require considerable revision if they are to be of any use. The view of causation that follows is an attempt at such a revision.

Consider again the entity we have been taking as our example. We can hardly take the Person seriously as an entity unless we regard him as a center of causal efficacy. Indeed, taken merely as an example, the Person is a better model (before philosophical criticism) than some other naive models. The causal efficacy of a rolling billiard ball, for instance, is regressive for common sense in a way in which the action of a friend in a crisis, for all the forces bearing upon him, is not. And we are dissatisfied when a philosophical analysis in alliance with science attempts to make his supposed efficacy regressive by describing that supposed person as merely a node in a complex of forces. That is not what we mean by a person, we say. That is not, I should add, what we mean by *any* entity when we take it in the way I have been expounding. If we take a person seriously *as an entity*, we must take *him* seriously as a center of causal efficacy when he operates in the world about him; if we take *any* entity seriously *as an entity*, we must take *it* seriously as a center of causal efficacy in its own world.

Let us confine ourselves for the moment, however, to our example, the Person. Taking him seriously when he operates in the world about him means that we must also take him seriously in relation to the subordinate causal forces that meet *in* him and in some sense do indeed cause him to be what he is. If we do not regard *him* as exercising a reciprocal 'causal'

301

influence upon *them*, the causal efficacy we think we see when he operates in the world about him vanishes, and with it his status as entity.

Notice that in all this our recognition of causation rests upon our recognition of entities and Being. Thus the recognition that the subordinate causal forces *are* causal forces rests upon our recognition of them *as* subordinate entities and subordinate levels of being within the entity we are attending to. The very sense we are giving 'causation' in their case is derivative from our recognition of them *as* entities and levels of being; and so, of course, with the reciprocal 'causal' influence exercised by the supervening entity, the Person, although this influence is obviously different from that exercised by the subordinates in (partially) causing him to be what he is. Conceived of as an expression or exemplification of the concrete and general power of Being, this causal influence permeates the subordinates in a way that, from the point of view of *their* dominance by *it*, appears 'purposive.' The word leads at once to difficulties, but one may at least suggest that the causal efficacy of an entity upon its subordinates (as distinct from the causal efficacy exercised upon *it* by its subordinates) is an exercise of power and order that, although displaying itself in process and in spatial configuration, is not wholly clarified in terms of the efficacy of earlier configurations in producing later ones. On this view the efficacy of any entity (*qua* supervening) is in part non-extensive, but what is thus non-extensive manifests both order and power in a way for which the idea of purpose is an inadequate but sometimes helpful metaphor.

We have, then, two basic senses of causation, both of them derived from the recognition of entities and Being. Let us call the first sense *ontic power:* it is the power exercised by an entity by the very 'act' of being an entity, and in the case of a supervening entity it is therefore exercised *on* the subordinate entities, although exercised *on* them only in the sense that it is exercised *by means of* them. It is the mode of power upon which such notions as 'formal cause' and 'final cause' shed some illumination and some deep and daunting shadows as well. It is a sense of causation in which the derivative character of the notion 'causation' is recognized. The second we shall call *causal power*, seeing that it is closer to a common-sense acceptation of 'causation.' It is the power exercised *by* an entity in its world, a power which, in the case of subordinate entities, includes a contribution to the being of a supervening entity. It is derivative from ontic power because it is exercised only in virtue of an entity's status as an entity.[5]

It will of course follow that the causal power exercised by subordinate entities in contributing to the being of a supervening entity will be in one sense simultaneous with the latter. We may express this by saying that, for each state of organization of the subordinate entities and levels, there will be in the supervening entity a simultaneous state upon which it exercises causal power. Thus, a physical state of the brain might 'correspond' with (in the sense of exercising causal power upon) a simultaneous mental state, and both might 'correspond' with a simultaneous total state of the entity. We can make out a similar case for the simultaneity of the ontic power of a supervening entity with the subordinates upon which, and by means of which, it is exercised. In another sense, which depends upon the non-extensive side of ontic power, we may think of the supervening entity as 'before' the organization of its subordinates which it brings about and in which its ontic power is expressed. There will remain many important senses in which the cause precedes the effect. One of these was mentioned in my earlier concession that for special purposes we may think of an entity as a complex event. But, as we have been concerned only to correct what appears to be an excessive emphasis on temporal priority, we may perhaps neglect other such senses here.

From this point of view reductionist explanations of life simply neglect levels of ontic power that are higher than the ones that concern them. A reductionist explanation of a living thing is an attempt to reduce the laws prevalent in life to those prevalent in the context dealt with by physics and chemistry. These laws express the configurations that can (or should we say must?) be taken by such entities as molecules. The contribution of such entities in their configurations to the higher 'entities' they form would correspond to what I have called causal power; the theoretical representation of the laws prevalent up through the molecular level would in effect be a representation of the ontic power of those levels. To insist that there are also biological laws (non-reducible ones) is one way of

303

insisting that there are higher levels of ontic power. It is, however, by no means clear that higher levels of ontic power are best approached in terms of laws. Can we fully express the levels of ontic power exercised by a Person in terms of law?

III

We are very far from authenticating a metaphysical (or philosophical) knowledge of the sort I have described. Before turning to that topic we must try to distinguish this knowledge from other and more familiar kinds. Let me begin by introducing a figure whose influence has been in many ways anti-metaphysical: that of Kant.

Kant seems to have forbidden philosophical knowledge—he did so, at any rate, if one identifies philosophical knowledge with metaphysics—but at least he did direct the philosopher to interest himself not in knowledge of a scientific sort, and not in some supposed knowledge that is higher than scientific knowledge, but rather in the *framework* of all our knowledge. It was a framework that defined the nature and limits of what we might with justice call knowledge, and, as it was established by the structure of reason, philosophy's framework-concern consisted in the analysis of the structure of reason. In effect, then, Kant distinguished philosophy as concerned with the framework from the disciplines that work within the framework. As will appear if I succeed in making myself clear, I do not agree with Kant in many particulars, but I ask the reader to hold fast to the idea of philosophy as concerned with the framework. Although I intend something rather more concrete than the term 'framework' suggests, I shall contend that *philosophy is a mode of knowing that acts (or ought to act) as a framework for our other modes of knowing, and, by extension, as a framework for our various modes of experience.*

Kant occupied himself, to be sure, with the structure of an existing and as he thought unalterable framework. Because he defined knowledge largely in terms of the science of his day, and because the conceptions of that science did not deviate startlingly from common sense, it was a framework suited to scientific and common-sense activities. It was philosophy's concern to expound that framework, to show that it must

304

inevitably be the framework for anything that deserves to be called knowledge, and to show also why, on the level of cognition, we are not equipped to go beyond it. It was the very framework of the objectivity we find in the common-sense world and in science, but it was also subjective in origin, springing from the very character of our sensory powers, and to that extent a barrier to an apprehension of the thing-in-itself. It is not, I must emphasize, that kind of framework-concern that I am urging.

It is rather the *alterable* character of the framework for our various modes of knowledge, action, feeling, and experience that I want to point to. It seems to me that it should now be the proper task of philosophy to alter it—alter it, I mean, by making it more adequate to the vitality and resource of the reality that lies all around us and in ourselves, more adequate indeed to what we need, want, and ought to become. The production of a framework more adequate in this sense is part of what I mean by the phrase 'philosophical knowledge.' It is, however, so creative a task that I envision that is necessary to add that it is not a question of producing a new framework, and then subjecting it to some sort of critical scrutiny, as a scientist might produce a new theory or model and then put it to the test. The point is rather that the act of increasing the adequacy of our framework is also the act of deepening our philosophical awareness.

But although the word 'framework' is truly appropriate here, it does not, as I said earlier, carry a sufficient burden of concreteness. I am not suggesting merely that we should make more adequate the conceptual network we attempt to fling over that recalcitrant thing, reality. We possess our appropriate framework—the framework that should enframe all our cognitive efforts of a more limited import—when, with a reason whose dynamism is an ordinate composure of our passional nature, we possess in a manner adequate to its nature any being whatever—a molecule, a tree, a dancer, a child, a work of art, or that most ambiguous and difficult of entities, the Person. To possess an entity adequately with the mind is not to misprize it; not to resolve it reductively into levels of lesser status; and not to miss its importance, value, or uniqueness.

The making more adequate of our framework, then—the framework that is *really* there though we should deny it, as a

305

moral or artistic value may be present though we should fail to see it—is here understood as the deepening (and authentication) of our grasp of beings and Being.

I have already suggested some ways in which this is a framework knowledge in claiming that we do not build our knowledge of entities up out of components in which the conception 'entity' is lacking. The conception 'entity' *enframes* (from this point of view) activities in which our knowledge of it grows more detailed and activities in which we deal with it more abstractly. In this sense these activities presuppose the enframing knowledge, even though the latter is inadequate and wants improvement. So with the conception of causation we here put forward. We cannot adduce an experience of power antecedent to this grasp of ontic power: rather, the identification of instances of power (however abstract) presupposes this enframing cognitive grasp of power—even though this grasp is inadequate and wants improvement.

There is another sense in which this is a framework knowledge: the categories 'reason' and 'experience' are both of them fused and inseparable from the category 'entity.' We profess to be aware of, or experience, entities and causation; but we also profess to *know* these things and to embody our knowledge in articulated judgments that have generality to them. The judgments do not hold apart from the experience, and the experience is incomplete and mute except as its uniqueness and particularity culminate in general judgments. Any supposed experience, any judgment of an isolated reason, any rational mode of inquiry (like science) involving an empirical and a theoretic component that we might bring against this knowledge, falls before the paradigmatic character of the union of reason and experience that characterizes it—a paradigmatic *joint satisfaction* of the demands of reason and experience.

It is, however, hardly a usable paradigm until it is authenticated. It seems that we must suppose this knowledge incomplete and must complete it to authenticate it.

IV

At the outset I described this metaphysical (or philosophical) knowledge as normative and reflexive. It is also peculiarly concrete, not only in the obvious sense that it is concerned to find our most concrete 'subject-matter,' but also in the sense that it draws very deeply upon our resources as persons. There is no question, then, of an authentication that subjects our cognitive faculties to a dispassionate analysis to see what they can and cannot do. We are also concerned about the relation of those faculties to the rest of the structure of the Person, and our concern involves the whole structure of the Person. Our authentication must come about, if at all, as the result of a self-knowledge that is congruent with the metaphysical knowledge we seek to authenticate. The Person is no longer an example but the very center of our enterprise. From this circle we cannot, and ought not to try to, escape; the very importance of our enterprise consists in our accepting it.

Let us try to enlarge upon these difficult points. Consider first the impulse that moves us to seek in a knowledge of entities and Being a framework for all our knowing. We wish to use the framework to settle epistemological questions of a very fundamental nature, as, for instance, the 'relation' between reason and experience. The philosophic impulse as I have depicted it will not permit us to remain satisfied with philosophic foundations that rest upon the evidence of experience alone (in whatever sense we take 'experience') or upon the evidence of reason alone (in whatever sense we take 'reason').[6] In this sense the the character of an acceptable foundation is already laid down by the demands of the philosophic impulse. We take it for granted that the enjoyment of all the particularity, immediacy, and uniqueness of an ultimate paradigm of experience is inseparable from the enjoyment of the generality, clarity, and self-evidence of an ultimate paradigm of rationality. Thus, although our concern for foundations is by definition a concern for establishing our right to invoke a variety of empirical and rational criteria in a variety of suitable circumstances, we expect the very attainment of that foundation to consist in a paradigmatic union of an ideal empirical and an ideal rational satisfaction. That is what the philosophic impulse (as here depicted) demands.

But it is not just a demand for epistemological foundations that moves us. Philosophy is a passionate enterprise, and that in a sense deeper and more concrete than the sense, so illuminated for us by Polanyi, in which *all* knowledge is a passionate enterprise. There is simply *more* of our affective or passional structure involved in philosophizing than in, say, our search for ultimate physical laws, and this structure is involved in a way that is most crucial for us. Our nature is very completely employed in it; the impulse is *concerned* with the completeness of our nature; and the impulse is also a demand for completeness in our 'subject-matter.'

Plato's conception of philosophy offers us a convenient example. For him philosophy was a passionate enterprise issuing in a reasonable insight, but the reasonable insight was in turn thought to be productive of a new order or integration in the sphere of the passions. His way of putting the matter is vivid enough: philosophy was in the root sense of the term an erotic enterprise. The philosopher was moved not just by a wish to discern the order in things, but by a passionate wish to bring about in himself his own best order, and this was one reason for his calling the cognitive achievement aimed at a wisdom. Knowledge of the most important kind about the reality that lies all around us was also self-knowledge, where 'self-knowledge' did not mean what it so often means today— just an understanding of the 'facts' of one's particular self— but the reordering, integration, or transformation of that self as well.

The framework knowledge I have in mind is inseparable from this side of the philosophic impulse. The level of order in which philosophic rationality emerges to take possession of its most important 'subject-matter,' which, we now see, includes itself as well, makes use of and alters the passions in this way. The alteration may be simple and dramatic, and no less important for leaving us still with the daily problems of self-control in great or small things. It may consist only in a curious shift in the relation between the system of our passions, with its foundation in the system of our bodies, and our rationality, a shift which makes it impossible for us to contemplate the bodily and affective levels of order that support reason as systems of facts upon which reason can act externally. Reason appears instead in such an association with those so-called facts as to make status as a (reasonable) Person not a factual status. Reason in this sense becomes in its association with its supporting levels a guide or norm that qualifies what we might otherwise consider mere factuality. And the qualification is very intimate, for one does not then stand outside the pattern of his passions, changing the operation of a factual system as one might tinker with a faulty mechanical or administrative structure. The change is deeper and more momentous than this. One's 'control' of his own factual system takes the form of an alteration of his complete factuality. *You are no longer a wholly factual system. The category simply does not apply.* In this sense a mode of knowledge a person felt he *ought* to bring about may contribute to his status as Person.

But in so far as the impulse to attain its most important 'subject-matter' is the urge also to authenticate that grasp, it is not only the status of the passions and their relation to rationality that enters into that 'subject-matter.' The 'subject-matter' also includes the question whether rationality, so supported by its subordinate levels, is capable of grasping the 'subject-matter' it seems to want, and if so, *how.* That entity, the Person, including the level of his philosophic rationality, now turns out to have been the entity upon which our supposed philosophic knowledge must focus if we are to justify a philosophic knowledge of *any* entity. The Person is no longer to be taken as just an example. The philosophic impulse, for all that it is an attempt at self-integration, is also a reflexive search for an epistemological foundation. We cannot avoid this tantalizingly reflexive character of philosophy.[7]

I do not wish to press the point about self-integration too far —do not wish to urge the absurd claim, for instance, that self-integration is not possible without metaphysics. One negative point seems, however, sound enough: mistrust your capacity to recognize entities and levels within them, and you will probably mistrust your ontological status as a person. On the positive side, however, it must be insisted that there is a crucially normative component in the entity that concerns us, the Person. In the first place its ontological status is inseparable from its value status: we ought to be persons, and often are not. In the second place, if there is a mode of attention that we ought to bring to entities and Being—a mode of attention, then, that the Person ought to seek to perfect—then the lack of it may pose a threat to the ontological status of the Person.

V

We must now face directly the tormenting reflexivity of our subject—a reflexivity that qualifies the other features of philosophic knowledge we have been discussing. I have associated the very creativity of reason with our impulse to make our framework more adequate and satisfying. In the long run this means only that the prime concern of our creativity is its own character. And this means that we are not seeking merely to deepen our self-knowledge, but to deepen our understanding of the *nature* of self-knowledge as well.

Facing the reflexivity of our problem has perhaps a greater simplicity than we might imagine. It means, first of all, only taking our reflective capacity seriously and permitting it to do confidently what our habitual use of it tells us that it can do. In this reflective exercise we recognize that reason is not only creative *within* some accepted framework, but creative even in its own reflective capacity and, therefore, confidently creative *of* its framework. Creativity in the act of reflection, I suggest, makes reflection itself not just, as it were, the mirroring of what is already accomplished, but the vital growing edge of accomplishment. What I call for, then, is a creative reflection to uncover and confirm the creativity of reason and therefore to uncover and confirm the creativity of reflection too. If this should be possible, we can perhaps come to see and enjoy reason's creative right to several frameworks and its right to the constant perfecting of its most important framework—that of its own self-knowledge. Tormentingly reflexive, indeed, even to a friendly critic; and to a hostile one rather like begging the question, for all my terminology seems to presuppose that questionable entity, the Person, set in a reality congruent with it.

It is this task of *bringing about* a more adequate mode of knowledge that leads me to call the reflective resource I have in mind *creative reflection* or *radically originative* reflection. The point is at once excessively subtle and absurdly simple. We turn away, first of all, from all images of reflection in which reflection is conceived of as the redoubling of one fact in another, as when a mirror (one fact) reflects some object (another fact). We begin by questioning the very kind of thing with which the word reflection is inevitably associated. Of course there are

310

senses in which reflection is a kind of redoubling, as for instance when I might at one instant be aware of some event and then at a later instant reflect upon my awareness of that event. But there are other senses of reflection, and some at least are simultaneous with the primary event, as when for instance I am both conscious of some thing or event and conscious that I am conscious of it. Self-consciousness, which is of course never merely *self*-consciousness, is a simple if mysterious paradigm of the sense of reflection I have in mind. Everybody is familiar with it, but we hardly clarify it in terms of images and images of images or in terms of one conscious moment remembered in a later conscious moment.

Second, we see at the root of every act of reasonable intelligence a fusion of understanding and awareness that is very like the fusion of the general and the unique, of value and fact, that I am claiming to find in the framework called the Person. For on the one hand there is a general self-evidence—at once a discovery, recognition, and acknowledgment of a realm of norms, standards, criteria, or values—and on the other hand, fused with this, as what it both qualifies and grows out of, the very particularity and uniqueness of whatever it is that falls under our attention. Technically speaking, there is in our concrete apprehension of any entity whatever a fusion of the general and the particular, of the One and the Many, of what has (too abstractly) been called the rational and empirical. And it is precisely this fusion that introduces with regard both to the act of reason and to what that act apprehends a note of value that ends at last in our conceiving of the whole idea of 'mere' fact as highly abstract. *But we see all this, not with an empirical introspective vision, but with a creative or originative reflection that is perfectly congruent with what it sees, being in fact nothing more than the heightened self-consciousness of what it sees. Everything that has been said of every act of reasonable intelligence applies to the reflective perfection of intelligence as well.*

Third, we see the whole complex act of reason as a creative response to what is before it. Reason lifts its object into the receptivity of its awareness and completes this awareness in the articulateness and generality of symbolic discourse. Its creativity does not consist in imposing a conceptual frame *upon* an object of which it is already aware (although there are less basic

311

activities *within* the framework for which this easy opposition of the rational and the empirical is an adequate image); its creativity consists rather in the *production* of its own awareness and the completion of that in a general realm which defines the nature of value for us. This is another way of stating the earlier claim that reason *produces* its own framework. *Once again we must think of all this as said as well of the reflective act by means of which we assert it: creative reflection lifts its object (reason) into the receptivity of its awareness and completes this awareness in the articulateness and generality of symbolic discourse. Or we may say that reason in creative reflection recognizes reason as creatively reflective: and the reduplication consists not in separate acts but in the separate sentences in which we articulate this point.*

The point of all this tortuous exposition, then, is that the reasonable self, focusing its energies in the creatively reflective act, wins to a position in which it can enjoy its own autonomous responsibility. But the autonomous responsibility is not just an epistemological one. We do not, that is, simply enjoy a combined rational-empirical satisfaction that then enables us to employ both rational and empirical criteria with confidence in appropriate circumstances. That is, to be sure, part of the story. The autonomous responsibility is also, as it were, existential: the act of creative reflection enables us also to *live* as Persons in the sense that we are authenticating the existence of a self that is normative in nature and that, being so, can also be persuaded to dissolve. We should, however, miss the delicacy of what is at issue if we did not press on to say that the act of creative reflection itself is informed by an ontological expansion of the Person. Philosophy is a very concrete exercise. The centrality of the reflective effort remains none the less: reflection has, so to speak, not merely followed upon the primary act of reason, or upon the primary existence of the self (fact reflecting fact). It has rather led these primary acts, lifting the whole enterprise into a region in which it can be properly evaluated because our *power* of evaluation is only thus liberated. Or we may say, more simply, that a heightened *self*-consciousness heightens our appreciation of the utter irreducibility of consciousness and self-consciousness to the models or images we sometimes employ in talking about them. And that *this* irreducibility becomes the

312

paradigm for the irreducibility of *any* level we focus upon to *lower* levels.

We return, then, to the theme of the unity of Being and causation that we began with: it is the ontic power of the Person that makes itself felt as our philosophic knowledge unfolds. A full causal explanation of any level of order that is not to be reductive must include the ontic power of the level in question. In *this* case that ontic power reveals itself in the originative, creative— or, as an older terminology might have it, the '*a priori*' or 'innate'—aspect of expanding rationality.

REFERENCES

[1] The terminological wrangles about the Trinity in early Christianity had no doubt many lamentable effects. For philosophers one of the most troublesome legacies is the word 'substance,' for if the problem or problems we usually designate with that word are in themselves obscure, they are surely made more so because of the name history makes us give them. Aristotle's word that is usually rendered as 'substance' is of course '*ousia*.' In Greek terminology the mature Trinitarian doctrine is that God is one divine *ousia* in three divine *hypostases*, and this becomes in Latin the claim that God is one divine *substantia* in three divine *personae*. That this is an odd result is by now a commonplace. *Substantia*, while it is surely not the best rendering of *ousia*, is an exact rendering of *hypostasis* (that which stands under).

We are not, of course, dealing with a mere translator's caprice, for although the force of the Aristotelian expression '*ousia*' is blurred by the terms '*substantia*' and 'substance,' Aristotle himself has furnished by his own vocabulary some of the overtones that might tempt one to render '*ousia*' as 'substance.' He uses, for instance, the Greek word '*hypokeimenon*' (that which *lies* under), which is a near etymological equivalent of '*substantia*,' sometimes to refer to matter as something that persists through its changes and sometimes to refer to an individual entity—a man, for instance—as something permanent underlying the changes it undergoes. Our Latin-based 'subject' is the equivalent of this Greek term, and it of course often turns up in the literature in place of 'substance.' The word 'substance,' even after a battering at the hands of epistemologists quite sufficient to render it thoroughly ambiguous, carries still, along with whatever else it may mean, this force of permanence in change.

[2] The expressions have come back into prominence in existentialism, in neo-Thomism, in Whitehead, and in recent translations of Aristotle's *Metaphysics*, notably Owens' and Hope's, where they are used, sometimes with qualifiers, to render '*ousia*.'

[3] Although there are some important differences between these claims and the doctrine of Michael Polanyi—differences suggested here, perhaps, by my use of the over-simple expression 'self-justifying,' as well as by my

insistence on the word 'imperronal'—there is nevertheless a congruence between his distinction between subsidiary and focal awareness and my claims about the lack of equivalence between an assemblage of data and a judgment about entities.

⁴ The meeting of formal and final causation in Aristotle's doctrine of essence is a case in point. We suppose the entity already complete and, as it were, making itself by controlling the causal forces that enter into it. We need not trouble ourselves here with the difficulties of this old but by no means exhausted doctrine. What I wish to say now is that the doctrine in its usual form is an insufficiently subtle way of stating the point that to *be* an entity (in the fundamental sense we are seeking) is to be a mode of order in which contributory causal influences are in turn subject to some control. And, further, that if we run into paradoxes in elucidating that control either in terms of the sense of 'cause' used for the contributing forces or in terms of some other sense allied with the telic tradition, we may perhaps avoid the paradoxes by suggesting that it is the *being* of an entity that exercises that reciprocal control.

⁵ It is thus unprofitable to think of an entity as wholly *caused*, but unprofitable also to think of an entity as *self*-caused. To think of an entity as wholly caused is to forget that we derive our basic sense of cause from an entity's status *qua* entity. To think of an entity as self-caused is similarly to set the category 'cause' above that of 'entity,' however much it might seem intended to protect the freedom and integrity of the entity.

⁶ The dissatisfaction extends *a fortiori* to foundations that consist in our taking as basic some received attitude (such as common sense), or some received activity (such as discourse in ordinary language or the procedures of science) in which both empirical criteria and rational criteria of various sorts play their roles.

⁷ I am not, to repeat an earlier warning, claiming that in introspection we gain some particularly immediate knowledge from which we move by analogy to authenticate our knowledge of remote things. I am claiming simply that our *capacity to know* beings comes to assured functioning in and with our knowledge of ourselves as persons.

12

THE STRUCTURE
OF CONSCIOUSNESS

Michael Polanyi

Sir Francis Walshe, in whose honor this essay was first published,[1] has often spoken of the inadequacy of anatomic structures to account for the full range of mental actions; he insisted on the presence of integrative mental powers not explicable in these terms. Toward the end of this paper I shall give reasons supporting this view.

I. TWO KINDS OF AWARENESS

I shall start with an analysis of perception and shall arrive by successive generalizations of the result at a stratified structure of living things, which will include the structure of consciousness in higher animals.

Take a pair of stereoscopic photographs, viewed in the proper way, one eye looking at one, the other eye at the other. The objects appear then distributed in depth, more rounded and real, harder and more tangible. This result is due to slight differences between the two pictures, taken from two points a few inches apart. All the information to be revealed by the stereoscopic viewing is contained in these scarcely perceptible disparities. It should be possible to compute from them the spatial dimensions of the objects and their distribution in depth, and I could imagine cases in which the result of such processing may be of interest. But this would not tell us what the things photographed look like. If you want to remember a family party or identify a criminal, you must integrate the stereo-pictures by looking at them simultaneously with one eye on each.

When looking at the stereo-image, we do see the separate pictures too; for we see the stereo-image only because we have a precise impression of the two pictures which contribute to it.

315

But we must distinguish between the two kinds of seeing: we are *focusing our attention on the stereo-image*, while we *see the two pictures only as they bear on the stereo-image*. We don't look at these two in themselves, but see them as clues to their joint appearance in the stereo-image. It is their function to serve as clues.

We may describe the situation by saying that we are *focally aware* of the stereo-image, by being *subsidiarily aware* of the two separate pictures. And we may add that the characteristic feature of subsidiary awareness is to have a *function*, the function of bearing on something at the focus of our attention. Next we may observe that the focal image, into which the two subsidiary pictures are fused, *brings out their joint meaning*; and thirdly, that this fusion *brings about a quality* not present in the appearance of the subsidiaries. We may recognize then these three features as parts of a process of knowing a focal object, by attending subsidiarily to the clues that bear on it. We meet here the structure of *tacit knowing*, with its characteristic *functional*, *semantic*, and *phenomenal* aspects.

I have developed this analysis of tacit knowing many times before and have now chosen the example of stereoscopic viewing, in order to prevent a recurrent misconception.[2] It is a mistake to identify subsidiary awareness with subconscious or pre-conscious awareness, or with the fringe of consciousness described by William James. The relation of clues to that which they indicate is a *logical relation* similar to that which a premise has to the inferences drawn from it, but with the important difference that tacit inferences drawn from clues are not explicit. They are informal, tacit.

Remember that Helmholtz tried to interpret perception as a process of inference, but that this was rejected, because optical illusions are not destroyed by demonstrating their falsity. Tacit inference is like this. The fusion of the two stereoscopic pictures to a single spatial image is not the outcome of an argument; and if its result is illusory, as it can well be, it will not be shaken by argument. The fusion of the clues to the image on which they bear is *not a deduction* but an *integration*.

Jean Piaget has drawn a striking distinction between a senso-rimotor act and an explicit inference. Explicit inference is reversible: we can go back to its premises and go forward again to its conclusions, rehearse the whole process as often as we like.

316

This is not true for the sensorimotor act: for example, once we have seen through a puzzle, we cannot return to an ignorance of its solution.

The seeing of two stereo-pictures as one spatial image is, indeed, irreversible in two senses. Firstly, it is difficult to find our way back to the clues in the two pictures, because they are hardly visible. And there are many other clues to seeing something, like memories and the feeling inside our eye muscles, which we either cannot trace or cannot experience in themselves at all; they are *largely submerged, unspecifiable.*

Secondly—and this is more important for us—to go back to the premises of a tacit inference brings about its reversal. It is not to retrace our steps, but to efface them. Suppose we take out the stereo-pictures from the viewer and look at them with both eyes. All the effects of the integration are cancelled; the two pictures no longer function as clues, their joint meaning has vanished. What has happened here may be regarded as the inverse of tacit inference; a process of *logical disintegration has reduced a comprehensive entity to its relatively meaningless fragments.*

The best-known example of this is the way a spoken word loses its meaning if we repeat it a number of times, while carefully attending to the movement of our lips and tongue and to the sound we are making. All these elements are meaningful, so long as we attend through them to that on which they bear, but lose their meaning when we attend to them in themselves, focally. The famous tight-rope walker, Blondin, says in his memoirs that he would instantly lose his balance if he thought directly of keeping it; he must force himself to think only of the way he would eventually descend from the rope.[3] When flying by aeroplane first started, the traces of ancient sites were revealed in fields over which generations of country folk had walked without noticing them. Once landed, the airman could no longer see them either.

The purpose of this paper is to show that the relation between body and mind has the same logical structure as the relation between clues and the image to which the clues are pointing. I believe that the paradoxes of the body-mind relation can be traced to this logical structure and their solution be found in the light of this interpretation.

The example of stereo-vision stands of course for a wide

range of similar intellectual and practical feats of knowing. We know a comprehensive whole, for example a dog, by relying on our awareness of its parts for attending focally to the whole. When we perform a skill, we attend focally to its outcome, while being aware subsidiarily of the several moves we co-ordinate to this effect. I have carried out this analysis often elsewhere and shall take it for granted here.[4] But there is a further step which I must restate once more. I shall say that we observe external objects by being subsidiarily aware of the impact they make on our body and of the responses our body makes to them. All our conscious transactions with the world involve our subsidiary use of our body. And our body is the only aggregate of things of which we are aware almost exclusively in such a subsidiary manner.

I am speaking here of *active* consciousness, which excludes incoherent dreams or pathological bursts of temper. Active consciousness achieves coherence by integrating clues to the things on which they bear or integrating parts to the wholes they form. This brings forth *the two levels of awareness*: the lower one for the clues, the parts or other subsidiary elements and the higher one for the focally apprehended comprehensive entity to which these elements point. A deliberate act of consciousness has therefore not only an identifiable object as its focal point, but also a set of subsidiary roots which function as clues to its object or as parts of it.

This is the point at which our body is related to our mind. As our sense organs, our nerves and brain, our muscles and memories, serve to implement our conscious intention, our awareness of them enters subsidiarily into the comprehensive entity which forms the focus of our attention. A suitable term is needed to speak of this relation briefly. I shall say that we attend *from* the subsidiary particulars *to* their joint focus. Acts of consciousness are then not only conscious *of* something, but also conscious *from* certain things which include our body. When we examine a human body engaged in conscious action, we meet no traces of consciousness in its organs; and this can be understood now in the sense that subsidiary elements, like the bodily organs engaged in conscious action, lose their functional appearance when we cease to look *from* them at the focus on which they bear, and look instead *at* them, in themselves.

The way we know a comprehensive entity by relying on our awareness of its parts for attending to its whole is the way we are aware of our body for attending to an external event. We may say therefore that we know a comprehensive entity by *interiorizing* its parts or by making ourselves *dwell in them*; and the opposite process of switching attention to the parts can be described as turning the parts into *external objects* without functional meaning; it is to *externalize* them.

This formulation of tacit knowing is particularly suited for describing the way we know another person's mind. We know a chess-player's mind by dwelling in the stratagems of his games and know another man's pain by dwelling in his face distorted by suffering. And we may conclude that the opposite process, namely of insisting to look at the parts of an observed behavior as several objects, must make us lose sight of the mind in control of a person's behavior.

But what should we think then of current schools of psychology which claim that they replace the study of mental processes by observing the several particulars of behavior as objects and by establishing experimentally the laws of their occurrence? We may doubt that the identification of the particulars is feasible as they will include many unspecifiable clues; but the feasibility of the program will not only be uncertain, but logically impossible. To objectivize the parts of conscious behavior must make us lose sight of the mind and dissolve the very image of a coherent behavior.

Admittedly, behaviorist studies do not reach this logical consequence of their program. This is due to the fact that we cannot wholly shift our attention to the fragments of a conscious behavior. When we quote a subject's report on a mental experience in place of referring to this experience, this leaves our knowledge of that experience untouched; the report has in fact no meaning, except by bearing on this experience. An experimenter may speak of an electric shock as an objective fact, but he administers it only because he knows its painful effect. Afterward he observes changes in the conductivity of the subject's skin which in themselves would be meaningless, for they actually signify the expectation of an electric shock—the skin response is in fact but a variant of goose-flesh.

Thus a behaviorist analysis merely paraphrases mentalist

descriptions in terms known to be symptoms of mental states and its meaning consists in its mentalist connotations. The practice of such paraphrasing might be harmless and sometimes even appropriate, but a preference for tangible terms of description will often be restrictive and misleading. The behaviorist analysis of learning, for example, has banned the physiognomies of surprise, puzzlement, and concentrated attention, by which Koehler described the mental efforts of his chimpanzees. It avoids the complex, delicately graded situations which evoke these mental states. The study of learning is thus cut down to its crudest form, known as conditioning. And this oversimple paradigm of learning may then be misdescribed as it was by Pavlov, when he identified *eating* with an *expectation to be fed*, because both of these induce the secretion of saliva. Wherever we define mental processes by objectivist circumlocutions, we are apt to stumble into such absurdities.

The actual working of behaviorism confirms, therefore, my conclusion that strictly isolated pieces of behavior are meaningless fragments, not identifiable as parts of behavior. Behaviorist psychology depends on covertly alluding to the mental states which it sets out to eliminate.

II. PRINCIPLES OF BOUNDARY CONTROL

But is not the material substance of all higher entities governed throughout by the laws of inanimate matter? Does it not follow then that it must be possible to represent all their workings in terms of these laws? Yes, this would follow. If I claim that these higher entities are irreducible, I must show that they are governed in part by principles beyond the scope of physics and chemistry. I shall do so. I shall show first that a number of different principles can control a comprehensive entity at different levels. I have repeatedly presented this theory in more particular terms.[5] It will be developed here on general lines.

There exist principles that apply to a variety of circumstances. They can be laws of nature, like the laws of mechanics, or be principles of operation, like those of physiology, as for example those controlling muscular contraction and co-ordination; or they can be principles laid down for the use of artifacts, like the vocabulary of the English language or the rules of chess. Not all important principles have such wide scope; but I need not go into this, for it is enough to have pointed out that some principles exist that do.

We can go on to note then that such a principle is necessarily compatible with any restriction we may choose to impose on the situation to which it is to apply; it leaves wide open the conditions under which it can be made to operate. Consequently, these conditions lie beyond the control of our principle and may be said to form its boundaries, or more precisely its *boundary conditions*. The term 'boundary conditions'—borrowed from physics—will be used here in this sense.

Next we recognize that in certain cases the boundary conditions of a principle are in fact subject to control by other principles. These I will call higher principles. Thus the boundary conditions of the laws of mechanics may be controlled by the operational principles which define a machine; the boundary conditions of muscular action may be controlled by a pattern of purposive behavior, like that of going for a walk; the boundary conditions of a vocabulary are usually controlled by the rules of grammar, and the conditions left open by the rules of chess are controlled by the stratagems of the players. And so we find that machines, purposive actions, grammatical sentences, and games of chess are all entities subject to *dual control*.

Such is the stratified structure of comprehensive entities. They embody a combination of two principles, a higher and a lower. Smash up a machine, utter words at random, or make chess moves without a purpose, and the corresponding higher principle—that which constitutes the machine, that which makes words into sentences, and that which makes moves of chess into a game—will all vanish and the comprehensive entity which they controlled will cease to exist.

But the lower principles, the boundary conditions of which the now effaced higher principles had controlled, remain in operation. The laws of mechanics, the vocabulary sanctioned by the dictionary, the rules of chess, they will all continue to apply as before. Hence no description of a comprehensive entity in the light of its lower principles can ever reveal the operation of its higher principles. *The higher principles which characterize a comprehensive entity cannot be defined in terms of the laws that apply to its parts in themselves.*

321

On the other hand, a machine does rely for its working on the laws of mechanics; a purposive motoric action, like going for a walk, relies on the operations of the muscular system which it directs, and so on. The operations of higher principles rely quite generally on the action of the laws governing lower levels.

Yet, since the laws of the lower level will go on operating, whether the higher principles continue to be in working order or not, the action of the lower laws may well disrupt the working of the higher principles and destroy the comprehensive entity controlled by them.

Such is the mechanism of a two-levelled comprehensive entity. Let me show now that the two-levelled logic of tacit knowing performs exactly what is needed for understanding this mechanism.

Tacit knowing integrates the particulars of a comprehensive entity and makes us see them forming the entity. This integration recognizes the higher principle at work on the boundary conditions left open by the lower principle, by mentally performing the workings of the higher principle. It thus brings about the *functional structure* of tacit knowing. It also makes it clear to us how the comprehensive entity works: it reveals the meaning of its parts. We have here the *semantic aspect* of tacit knowing. And since a comprehensive entity is controlled as a whole by a higher principle than the one which controls its isolated parts, the entity will look different from an aggregate of its parts. Its higher principle will endow it with a stability and power appearing in its shape and motions and usually produce also additional novel features. We have here the *phenomenal aspect* of tacit knowing.

And finally, we are presented also with an ontological counterpart of the *logical disintegration* caused by switching our attention from the integrating center of a comprehensive entity to its particulars. To turn our attention from the actions of the higher principle, which defines the two-levelled entity, and direct it to the lower principle controlling the isolated parts of the entity, is to lose sight of the higher principle and indeed of the whole entity controlled by it. The logical structure of tacit knowing covers thus in every detail the ontological structure of a combined pair of levels.

322

III. APPLICATION OF THESE PRINCIPLES TO MIND AND BODY

The next question is whether the functioning of living beings and of their consciousness is in fact stratified. Is it subject to the joint control of different principles working at consecutive levels?

The laws of physics and chemistry do not ascribe consciousness to any process controlled by them; the presence of consciousness proves, therefore, that other principles than those of inanimate matter participate in the conscious operations of living things.

There are two other fundamental principles of biology which are beyond the scope of physics and chemistry. The structure and functioning of an organism is determined, like that of a machine, by constructional and operational principles which control boundary conditions left open by physics and chemistry. We may call this a *structural principle*, lying beyond the realm of physics and chemistry. I have explained this a number of times before and will not argue it here again.[6]

Other functions of the organism not covered by physics and chemistry are exemplified by the working of the morphogenic field. Its principles are expressed most clearly by C. H. Waddington's 'epigenetic landscapes'. These show that the development of the embryo is controlled by the gradient of potential shapes, in the way the motion of a heavy body is controlled by the gradient of potential energy.[7] We may call this principle an *organizing field* or speak of it as an *organismic principle*.

Most biologists would declare that both the principles of structure and of organizing fields will be reduced one day to the laws of physics and chemistry. But I am unable to discover the grounds—or even understand the meaning—of such assurances, and hence I will disregard them and recognize these two principles as actually used in biology today.

Living beings consist in a hierarchy of levels, each level having its own structural and organismic principles. On the mental level, explicit inferences represent the operations of fixed mental structures, while in tacit knowing we meet the integrating powers of the mind. In all our conscious thoughts these two modes mutually rely on each other and it is plausible to assume that explicit mental operations are based on fixed neural networks, while tacit integrations are grounded mainly in organizing fields. I shall assume also that these two principles are interwoven in the body, as their counterparts are in thought.

The purpose of this paper is to explain the relation between body and mind as an instance of the relation between the subsidiary and the focal in tacit knowledge. The fact that any subsidiary element loses its meaning when we focus our attention on it explains the fact that when examining the body in conscious action, we meet no traces of consciousness in its organs. We are now ready to complete this project.

We have seen that we can know another person's mind by dwelling in his physiognomy and behavior; we lose sight of his mind only when we focus our attention on these bodily workings and thus convert them into mere objects. But a neurophysiologist, observing the events that take place in the eyes and brain of a seeing man, would invariably fail to see in these neural events what the man himself sees by them. We must ask why the neurologist cannot dwell in these bodily events, as he could in the subject's physiognomy or intelligent behavior.

We may notice that the latter kind of indwelling, for which we appear to be equipped by nature, enables us to read only *tacit* thoughts of another mind: thoughts and feelings of the kind that we may suitably ascribe to organismic processes in the nervous system. We can get to know the *explicit* thoughts of a person—which correspond to anatomically fixed functions of the nervous system—only from the person's verbal utterances. The meaning of such utterances is artificial; though ultimately based on demonstrations pointing at tacit experiences, such utterances have no direct appeal on the native mind. The facility for indwelling can be seen to vary also when prehistoric sites, unperceived from the ground, are discerned from the air. Our incapacity for experiencing the neural processes of another person in the manner he experiences them himself may be aligned with these gradual variations of indwelling.

We arrive thus at the following outline. Our capacity for conducting and experiencing the conscious operations of our body, including that of our nervous system, lies in the fact that we dwell fully in them. No one but ourselves can dwell in our body directly and know fully all its conscious operations; but our consciousness can be experienced also by others to the extent to which they can dwell in the external workings of our mind from outside. They can do this fairly effectively for many tacit workings of our mind by dwelling in our physiognomy and behavior; such powers of indwelling are fundamentally innate in us. By contrast, our explicit thoughts can be known to others only by dwelling in our pronouncements, the making and understanding of which is founded on artificial conventions. Objectivization, whether of another person's gestures or of his utterances, cancels our dwelling in them, destroys their meaning and cuts off communication through them. The nervous system, as observed by the neurophysiologist, is always objectivized and can convey its meaning to the observer only indirectly, by pointing at a behavior or at reports that we understand by indwelling.

The logic of tacit knowing and the ontological principles of stratified entities were derived here independently of each other, and we found that our tacit logic enables us to understand stratified entities. It shows us then that the higher principle of a stratified entity can be apprehended only by our dwelling in the boundary conditions of a lower principle on which the higher principle operates. Such indwelling is logically incompatible with fixing our attention on the laws governing the lower level. Applied to mind and body, as to two strata in which the higher principles of the mind rely for their operations on the lower principles of physiology, we arrive at three conclusions.

1. No observations of physiology can make us apprehend the operations of the mind. Both the mechanisms and organismic processes of physiology, when observed as such, will ever be found to work insentiently.

2. At the same time, the operations of the mind will never be found to interfere with the principles of physiology, nor with the even lower principles of physics and chemistry on which they rely.

3. But as the operations of the mind rely on the services of lower bodily principles, the mind can be disturbed by adverse changes in the body, or be offered new opportunities by favorable changes of its bodily basis.

The way integration functions in tacit knowing, as well as the presence of irreducible organismic principles in living beings, are both consonant with the arguments presented by Sir

325

Francis Walshe for the presence of integrative mental powers, not accounted for by the fixed anatomic structures of the central nervous system.[8]

IV. RETROSPECT

Many philosophic efforts of our century can be seen to have pointed toward such conclusions. A systematic attempt to safeguard the content of unsophisticated experience against the effects of a destructive analysis was made by Edmund Husserl during the first three decades of this century with far-reaching influence on continental philosophy. But its bearing on the body-mind problem was derived mainly later by Merleau-Ponty in his *Phenoménologie de la Perception* (1945). He gives a vivid and elaborate description of the way we experience our body. The body is 'known to us,' he writes, 'through its functional value'; its parts engaged in the performance of our actions 'are available to us in virtue of their common meaning';[9] our body expresses meaning but 'language does not express thought, it *is* the subject's taking up of a position in the world of his meanings.'[10] 'If a being is conscious it must be nothing but a network of intentions';[11] 'I do not understand the gestures of others by an act of intellectual interpretations. . . . The act by which I lend myself to the spectacles must be recognized as irreducible to anything else';[12] our experience of our body is an existential act, not based either on observation or on explicit thought. These remarks foreshadow my analysis, but I find among them neither the logic of tacit knowing nor the theory of ontological stratification, which I regard as indispensable for the understanding of the phenomena described by Merleau-Ponty.

Another follower of Husserl, Dr. F. S. Rothschild, arrived even earlier at the conclusion that the mind is the meaning of the body.[13] He developed this idea widely in neurophysiology and psychiatry, where I am not competent to follow him.

The mainstream of contemporary English and American philosophy ignores the inquiries of phenomenologists. But it shares their rejection of Cartesian dualism, and the kinship of the two movements goes beyond this. Deprive my quotations from the *Phenomenology of Perception* of their existentialist perspective, and they can be equated with observations of Ryle in the *Concept of Mind* (1949).[14] But such a transition brings out

326

the theoretical inadequacy of these descriptions and results in drawing false conclusions from them. Take a simple example. Merleau-Ponty says 'I do not understand the gestures of others by an act of intellectual interpretation' and Ryle says the same: 'I am not inferring to the workings of your mind, I am following them';[15] but Merleau-Ponty finds an alternative to 'intellectual interpretation' in existential experience, while Ryle has none and affirms, therefore, that 'most intelligent performances are not clues to the mind; they are those workings,'[16] which is absurd. Many vivid and often subtle phenomenological descriptions are used by Ryle to demonstrate that the mind does not explicitly operate on the body, and from this result he concludes that body and mind are 'not two things,'[17] 'not tandem operations,'[18] containing no 'occult causes,'[19] 'no occult antecedents,'[20] no 'ghost in the machine,'[21] in other words, no Cartesian duality. But what actually follows from the fact that mind and body do not interact explicitly is that they interact according to the logic of tacit knowing. And it follows further that this logic disposes of the Cartesian dilemma by acknowledging two mutually exclusive ways of being aware of our body.

As Ryle's powerful argument leads him to fallacious conclusions, it offers a compelling demonstration of the troubles arising from the absence of the cognitive and ontological principles outlined in the present paper; that is why I selected his work for representing anti-Cartesian thought in contemporary British and American literature.

REFERENCES

[1] *Brain, 88* (1965), pp. 799-810.
[2] Recent publications of the author on which this paper draws: 'Clues to an Understanding of Mind and Body,' *The Scientist Speculates* (I. J. Good, ed.), London: Heinemann, 1962, p. 67; 'Tacit Knowing and Its Bearing on Some Problems of Philosophy,' *Reviews of Modern Physics, 34* (1962), pp. 601-16; 'Science and Man's Place in the Universe,' in *Science as a Cultural Force* (H. Woolf, ed.), Baltimore: Johns Hopkins Press, 1964, O.U.P., 1965; 'On the Modern Mind,' *Encounter* (May, 1965); 'The Logic of Tacit Inference,' *Philosophy* (Jan. 1966) pp. 1-18; 'The Creative Imagination,' *Tri-Quarterly* (1966); *The Tacit Dimension*, Garden City: Doubleday, 1966.
[3] Referred to in F. J. J. Buytendijk, *Traité de Psychologie Animale*, Paris: Presses Universitaires de France, 1952, p. 126.

[4] See 2, above.

[5] *Ibid.*

[6] *Ibid.*

[7] Cf. e.g., C. H. Waddington, *The Strategy of the Genes*, London: Allen & Unwin, 1957; particularly the explanation of genetic assimilation, p. 167.

[8] F. Walshe, *Critical Studies in Neurology* and *Further Critical Studies in Neurology with other Essays and Addresses*, Edinburgh: Livingstone and Co., 1948 and 1965, respectively.

[9] M. Merleau-Ponty, *Phenomenology of Perception*, London: Routledge, 1962, p. 149.

[10] *Ibid.*, p. 193.

[11] *Ibid.*, p. 121.

[12] *Ibid.*, p. 185.

[13] See Rothschild's earlier writings, which extend back to 1930. A fairly recent summary of them is given in the monograph: F. S. Rothschild, *Das Zentralnervensystem als Symbol des Erlebens*, Basel and New York: S. Karger, 1958, VII, pp. 1–134. In this monograph, Dr. Rothschild points out on pp. 10–11 that the meaning of the CNS manifested in consciousness is lost by examining the CNS as an object—just as the denotative meaning of a word is lost by such an examination. This anticipates part of my theory of body and mind. For a briefer summary, in English, of Dr. Rothschild's work, see F. S. Rothschild, 'Laws of Symbolic Mediation in the Dynamics of Self and Personality,' *Annals of the New York Academy of Sciences, 96* (1962), pp. 774–84.

[14] G. Ryle, *Concept of Mind*, London: Hutchinson, 1949.

[15] *Ibid.*, p. 61.

[16] *Ibid.*, p. 58.

[17] *Ibid.*, p. 74.

[18] *Ibid.*, p. 46.

[19] *Ibid.*, p. 50.

[20] *Ibid.*, p. 115.

[21] *Ibid.*, pp. 15–16.

EPILOGUE

COSMOS AND KINGDOM

Elizabeth Sewell

Five Stations, with end-pieces, on a chapter from *Heinrich von Ofterdingen*

I. TO FRIEDRICH VON HARDENBERG, THE BELOVED NOVALIS

To you, then, this third time.

For a rescue from abstract language
(Though poetry seem to us all a feint or illusion)
For a figure of what we are seeking.

Put it this way: there are men on a venture.
After? one could say they are after cosmos and kingdom.
Where? where else but the living rock-vaults of mind and body?

To you, again, for an image.

The first time it was to see the patterning
(This in *Die Lehrlinge zu Sais*)
                                        of the cosmos,
The great structures of things,
The ideas of the universe
                        in their several speech
Of crystals, clouds, wind-flaws over water,
Of mathematics, philosophy, poetry, natural science
(You studied them all, at Leipzig, Jena, Freiburg)
Which, themselves answering patterns of a universe,
Go towering into dreams,
Red rock crevasses split to the primal diorite,
Turquoise-translucent shells, parded skin of beasts,

331

The tabarded heraldry of all that lives and
Cries radiance to the morning sun;
These, with the strung senses and the attending heart,
Glyphs, numbers, symbols, words, figures,
Are working languages for scientist taxonomer
                        and poet taxonomer
(Languages to which this twenty-year-old would add
Music, erotic love, folktales and the Word Incarnate)
Languages which all, each to his own, must live and search by
If the world will be ordered and made.
Cosmos was here, God knows, without and within.
Kingdom also? Ivanov hints at it,
Exchanging those letters with his political opponent,
                        his fellow-patient,
Each in opposite corners of the hospital ward,
Steel beds
        the vacuum of sickness
                the revolution outside in the streets,
And the poet writing, there:
'We are all of us apprentices at Sais.'
(There? those innocent gatherers of flowers, stones, feathers,
To lay them out in rays, harlequin diagrams, thought-runes
                    of stars and divinations?)
There indeed
In 1917.

To you, on an instinct.

The second time was to learn interpreting
(This in *Hymnen an die Nacht*)
                        by one's own life,
And was conducted in darkness
To the sound of waters,
The flesh slumbering like a cloud on a high hill.
Methods beyond analysis, of metaphor, flashfire
Shimmers between two poles.
This is a thinking first, that is, poetry,
Peculiar energy of synthesis, spontaneity illuminating
The deeps of the world,
As the fireflies in their elixir

Of absolute forest gloom and damp and scent
From the guessed-at exquisite sugarpink massing sprays of
mountain laurel,
As those crazier insects,
cosmic dragonflies,
The comets whirring
Over the lake of darkness.
A thinking first, but that implicates
The life as well,
Waking or dreaming—*do I wake or dream?*
The man become metaphor (your word, Novalis)
Each man alive, to himself and others a part of the process
By which the cosmos is grasped,
The life become poem (in Milton or Goethe)
The life that is allegory (in Keats and Coleridge);
For the beloved shadows assemble here
Under the faint groves,
Naked all, as in Eden,
And murmur of power, a twinned ambiguous murmur,
Two-sided as
history and natural history
nature and human nature
body and body politic
these are numinous twins,
Two-sided as, where they walk, the scarcely moving leaves,
Tongues whether olive or silver,
Dappling in the no-light and
Mystery of the terrible stars
Which we must read, and read by,
As poets by lives of fire and myth we spin behind us;
Take into cosmos and kingdom this way of thinking
of anguish and
struggle
of loving and dying
the poets enjoin

Knitting us by our own nature into Nature
Knitting us by our own histories into History
Knitting us by our own organs into the *Polis*

To you, now, for this third and harder work.

Back, a century and a half? To unanimous voices,
Back, beyond faculties split and follies divisive,
To a cluster of poets (the scientists take it up later):

The organic whole, mind-body, imagining reason—

Each, according to his own stamp, they witness—

    The first
    Building in ice and sunlight those airy singing domes,
    Vision nascent
                cosmic palace
                        the process of self-construction

    The second
    Climbing through mist up mountains, the water-thundering
                        landscape,
    To front with the forehead at last the serried constellations,
    Exalted kindred
                intellect with its figures
                      a bright and ample
                            power

    The third
    Sees in the three planes of time a city, forged and woven,
    And starry works of wheels that turn in the heavens,
    And prophesies of our thinking, unbodied, imageless,
                        spectral,
    The city will grow redhot, the looms with their vestures of
    Gold and sea-shot tissue ravel and flap
    In the burning winds, and the stars grind us

    The fourth
    Running towards us shouting, 'To imagine that which we
                      know,
    To act that which we imagine,' to be flung from us, caught
    In an elemental turmoil of air and water and fire,
    The toss and whirl of dreadful oceanic rainbows, with
                      Homer
    In his salt-sodden pocket

The fifth and dearest
Burning his life out in love and fever and effort,
Intense towards beauty neither belied nor answered
By soft showers over the heath, and the birds singing,
Till in the last assay
The dark eclipsing disc of death reveals the unmistakable
                                                    steadfast
Corona of heroic fire

To you, Novalis, next; most like, and different.

Passionate, young, many-crafted, your marvellous studies,
Mineralogy, chemistry—I set them down in no order but
Cosmos—geology, physics, mathematics—and kingdom—
History, law, administration, seeming no more than
    A fountaining energy's overflow

In your swiftness for love, transcendence, God, finding your
                                            twenty-eight years
    Too long and too slow

Most dear, most honoured, a vital friendship annulling that
                                            premature death
    A long time ago

I come to you believing there is freedom in both these realms
Of cosmos and kingdom
    natural and man-made relations
    science and history
                        for poet and thinker, who is every man,
    To move to and fro

(That science and poetry lie in each other's arms for ever
    We both of us know)

Forward from here: philosophy? *Blütenstaub*, pollen, you say,
To dust our antennae
    Or set us to grow
335

Method? body the mind and its ways (us all) in figures
    And study them so

And history?
        A cave.
            Oneself.
                Into which we go.

II. THE FIRST STATION: *They gather at the mouth of the cave*

The old miner:
        Why did I bring this company here tonight?
        Out of the convivial inn where we had met,
        As it seemed, by chance, these other travellers
        From farther north—merchants, the silent boy
        Who dreams and listens; I, the stranger here
        By origin and profession; and these countrymen,
        One of whom brings, I see, his little son.
        I wanted truth for them,
        And for myself, no one occasion lost
        To look into the organic rock, my mystery,
        Tendrils of ore, leaf metal, the rough buds
        Of crystal and amethyst, and their chemic springs.
        Real mystery begins with expert knowledge.
        Mine is deep, secret. I would impart it.
                        For the moment, though,
        What a lovely night . . .

The countrymen:
        There may be monsters in these caves, no less—
        And bones, they say—a treasure trove—a ghost—
        Every man hold his weapon in his fist—
        Poking and prying can be foolishness.

The merchants:
        Luck visits only the prudent and adroit
        And diligent, so here are ladder and hook
        And rope—but just the same, be with us, Luck!
        There might be something here we could exploit.

336

The child:
    Fifty-nine lovely stones under my bed,
    For which my father scolds me, calls me dunce—
    What a big cave—my torch burns white and red—
    Can we go in at once?

The poet:
    How active and clear men seem,
        Integral, unafraid!
        Is it then retrograde
      To be passionately aware,
      With the wide dark landscape there,
    Of nature's essential theme,
        Which is also the poet's trade:
    That only in something like a dream
        Are revelations made?

Dusk on the hills, the skies
    Are film and glass;
Moonlight like a language lies
Whose images philosophize
    The detailed blades of grass,
      Body of earth and air
        Nerved with responding strings,
      And a soft inverted glow lays bare
      The hidden seeds of things.

Coherent time is flawed
    To elemental dew,
    The very grains of light
Singled and sown abroad,
      Minuscule and august,
      Imagination's dust,
    With a past man never knew
    And a future whose review
In the dim and looming night
      Is wordlessly discussed.

The miner:
    The air in the caves is healthy. We'll go in.
    Torches? Yes. Five's enough. Chalk for a mark

Of our return route? Yes. Lastly, remember
We look for knowledge and experience,
Holy things both, as is the earth itself,
And every man. Our word is 'Keep together.'

The poet:
    The mind is full of gleams
        Of moon and fire
        On rocks' metallic wire,
           Flakes of mica and jade;
    Reason's meridian beams
        Are powerless to inspire,
           Though mandatorily obeyed;
    Flashes along the mountain seams
    Are mind's prodigious enthymemes
    And only in ecstasy and dreams
        Are revelations made.

III. THE SECOND STATION: *They halt in the second rock-chamber*

Here we emerge from the passage, again in a cavern
    Vaguer than the first, vaster . . .
Fear . . . Then control your breathing. Address your neighbour.

Light from our torch plumbs upward to no ceiling,
    As if we were out-of-doors,
A night of rock sounded by echoes only.

Bluntly our simplest say they are going back.
    Probably right. We all
Confront an irrelevant, compulsive terror.

Wait . . . Keep asking yourself what it is you fear—
    Darkness—enclosure—pressure—
Weight of a hill-range hooding our head and shoulders—

Delivered and locked into this mass, this body,
    Where beating about its den
The quivering spirit screams . . . —No more of that.

Listen: occasional crash of a drop on a stone;
    Look round: eyes can make out,
Where the glistening wall runs smooth, a fire-play of shadows.

Once drop the eyelids on caverns of skull, those phantoms
    Coalesce—are there two?
They stop our passage—they speak—of cosmos and kingdom—

O secular breath of the caves whose impulse curves
    Even the stalactite gestures,
Oracular sighing, great shades, we hear and attend you:

    Thus to depict the nature of mankind:
    A cave where men must sit, helpless, in gyves,
    The world's procession, and a fire, behind;
    They watch parades of shadows all their lives,
    Making crude sense of each as each arrives,
    Out of which image of imaginings—
    Does it not, Glaucon?—all they know derives.
    Out with our best, then, into the light of things,
    The lone bright eminence, the sunlit place of kings.

        Second I name the idols of the Cave:
        The individual man prone to illusion,
        Infinite error more infixed and grave
        By education, habit and infusion,
        Each several spirit in its dark reclusion
        (Plato his fable); let us then begone
        Into that light of nature whose effusion
        Knowledge and power restores, as once we shone,
        Clear as the cherubim and learn'd as Solomon.

But the heart, the heart replies: That is not the answer.
    And the fear is less already,
And we start to move—not flight but penetration—
    And apart, ahead, the voice of our leader cries:

    We have scanned the stars and studied their orbits,
    Plotting pathways, making prediction;
    Grown up in the ground-web of men's relations,
    Artifacts visible and invisible;

Now at this lower and innermost level,
Richer and darker and deeper-rooted,
I would have us handsel, hold and inherit
Another network, wide as the world—

Auricles, antres, the verberant hollows
Of middle earth and her antechambers,
Resonant and connected workings,
Crypts under continent, sea-sunk catacombs,
Which, riddling the world, await resolution;
Inlet and outlet for classical elements,
Air fluting or still in shafts and funnels,
Water sliding on tilted tables,
Underground lakes, the broad rain-terraces,
While on fringes and firths the great green ocean
Brims with music those murmurous throats,
And sunlight advances as far as it dare,
To dance reflections on roof and rock-wall,
Alveolar and mesmerist ripples;
Or chasms that plummet, past air and water,
Through honeycombed crust to the core of the matter,
The molten heart self-manifest
In fuliginous or passionate fire.

This is the labyrinth, this the image
Reticular I would have us remember
As we seek the self-unravelling clue—
A sense of place is but the beginning.

Consider now this cave with its contents
And scope and substance for observation,
Flora and fauna, what furniture else.
Spread out, and, sharing our scattered lights,
Shout to us all what you see—or hear.

'Bones—and big ones—and unfamiliar!'
        Bear it could be, hyena or bison.
'A bear has scratched all down the wall here!'
        Grazing graffiti of ice and water.
'There are milkwhite threads of mould or moss
On this rock—or a fungus!'

As fronding forms
Of calcite, gypsum, mineral dendrites.
'I have found a fossil—looks like a snail!'
Stone-into-flesh has played it backwards.
'Something swims in this runnel of water!'
'Under that sill what scuttles—a scorpion?'
'Squeak of a bat shot past my shoulder!'

Good! then gather your minds together,
Reimagine the global system
Which we have entered, wherein we stand,
Our nethermost and numinous network,
Now lend that image a sense of movement.

Up through the mineralogical *couches*,
Curded and cusped and cryptocrystal,
(They have their aeons of evolution,
But their petro-progress we shall pass over),
Only observe how colonization,
Unabated from the beginnings,
Carries forward a ceaseless current
Into cave and corner however clandestine,
And yet in collusion, a playing together;
Spore and sperm on salt and basalt,
Wet beasts and dry and the small web-wingers
Appropriating the avian air,
And all to animate earth's potential,
A serial process, and open-ended.

Finally, man, who chipped his flintheads,
Pursued his prey and made his magic,
Left painted masterpieces behind him,
Came to himself, learned to love.

All this is ourselves, my friends and brothers,
Here we are home, as under heaven,
And old as the hills we have our patent—
I would have us take up the sense of time.

We stand at the head of the moving column
. . . hovering . . .
                    shall we go on?

O let me speak to that!
I, the poet, who have stood here in the dark, silent,
Useless, it seemed, finding nothing.
You our guide
Who are also science and vision
Have given us
Reassurance, almost an exaltation,
Have given us,
As if for the first time,
Our eyes,
From the patient watching of detail to the great
Sweep and sympathy of the panorama,
Through bursts of sunshine and silverings of rain, down the
                                                    long years
Moving . . .

Let me now add to this gift
What is mine to give,
To the temporal senses and the interpreting mind
The wherewithal of earth's biography, which is
Music.

Not here, now now, that music of the spheres,
    Which man imagined, ringing, singing,
        And found the sound so ravishing,
        Perfect the more because inaudible,
            He matched it to his thoughts,
    Tempered the firmament with spun-glass stringing,
            Rays, chords, long speculations,
                Timing, chiming,
            And into the blue goes swinging
        The mind, to labour there until it disappears
        (Out of the sky what nebulous wrack of tears?)
    In the shimmering meshes of its own upbringing.

    But if one lay all day in a cleft
            Of the rocks' heart,
    Feeling earth shoulder over in musky sleep,
        A ground bass grows from under
        The purring of pulses and the sough of breath,

342

Inaudible only since we are part
Of that emerald mirk and thunder,
The breves and bells of pain, sudden cry of a rose,
Or far horns in the black ravine sounding to battle.
Once apprehend, wonder, and crown that strain:
Build honey-coloured cities, rooted deep,
Upon whose walls starred trumpeters the watches keep. . . . . .

. . . . . (Excuse me . . . I seem to be uttering archaic babble . . .)

Ahoy! This is Simon's boy! I want to report
From the end of the cave. Please, everyone, keep still!
One set of footprints leading into the dark,
And a man's voice singing deeper in the hill!

Who is still here? I the miner—the poet—
The merchants—well, commerce is part of the picture,
As education is—and the child awaits us.
As a band then, fellow-citizens, are we ready?
We shall go on now.

IV. THE THIRD STATION: *They enter the inner cave and meet its*
inhabitant

The merchants' tale:
As we came out from the long long walls
That were both cold and damp,
We came into the highest hall
Was lighted by a lamp.

As we came out of the ragged ways
And the groping hand to hand,
In the heart of the hill we found a man
Who greeted us like a friend.

As we came out and stood amazed,
We saw him sit so free,
With a slab of stone for table-top,
And a book upon his knee.

He was neither young nor old
    But silvered as the snow,
And nobly looked his countenance,
    And cheerfully also.

Sandals he wore upon his feet,
    And over all a cloak,
A gracious man to be a lord,
    And graciously he spoke.

'Now welcome, welcome be to me,
    The first that here have won.
Fair is the house of the good green world,
    And fair are moon and sun,
But fairest of all the things on earth
    Is the bonny face of man.'

'What do you here, most noble sir,
    If we may make so bold?
How came you to this lonely spot,
    For the ways are dark and cold?
How hangs on the wall that coat of mail,
    Embossed with beaten gold?'

'Only for quiet am I here,
    And the meditative mind.
I am no hater of the world
    Nor one who shuns mankind,

'Warrior from my early youth
    In many a tough campaign,
Heir to one of the noblest lines
    Come down from Charlemagne.

'Far have I travelled east and west,
    And held the reins of power,
And I have known the ways of love
    And had my heart's desire.'

'Your hermit's cave has other gear
    Which curiously we spy,
Book upon book piled on the floor,
    With a zither close thereby,
And graven in the table stone
    A man and woman lie.

'Two forms engarlanded with flowers
    That blossom to the life—
O what are the words stand written there,
    And had you e'er a wife?'

'Of that is dearest to his heart
    No mortal man can tell.
A wife I had, and two sweet babes,
    And I have buried all.

'This was the epitaph I made,
    Carven by mine own hand:
HERE IN THIS PLACE WE TWO RETURN
    UNTO OUR NATIVE LAND:
Wherefore my cheer and my presence here
    You haply understand.'

So we made an end of questioning
    In the high and haunted cave,
And some withdraw, some few approach
    A table, or a grave.

The man of action, the science man,
    The poet—there sit three,
One to listen, two to converse,
    And we stand back to see.

345

V. THE FOURTH STATION: *The conversation inside the hill*

The warrior-hermit   The miner-scientist
                    The poet (who does not speak aloud)

Where begin?   We began, for our part,
                                                  with the song. Warned
                                                  about wild beasts and,
                                                  worse, the preternatural,
                                                  we came through the
                                                  reaches of time and our
                                                  own thoughts to meet
                                                  your singing.

Begin then with love,
for of that the song
spoke, true love and a
serene looking towards
death; with love and
death, pinion and pole
of a life, my life, we
begin.

                    But we begin also
                    (silently I say it to
                    myself) as an emblem,
                    for the limner has cut
                    away the shell of the
                    hill and there we are,
                    printed, a crude wood
                    block, amidst what
                    text, the three men
                    sitting round a table
                    in the heart of a
                    hollow hill, to talk?

                                                  Yet there are other
                                                  things to your life,
                                                  if we divine by that:
                                                  this extreme retire-
                                                  ment, for instance,
                                                  which has brought you
                                                  into the kingdom that
                                                  I know, the mine
                                                  workings, faults,
                                                  impactions of the earth.

That came in stages;
I had when young a
passion to be a

solitary since only
so, thought I, could
the premonitions in
my mind be tracked
and come at.

There was talk once of
a well at the wood's
end, inexhaustible thus
through the winged
grasses to gaze and gaze.

You found yourself
in error?

I learned rather
that, first, heart
and mind and soul
must be flooded with
things seen and things
done, to flowing and
overflowing; that it
is the being among men,
the conviviality and
commitment, which in
the end gives selfhood
its true franchise.

Each age of a man may
have its natural
vocation: the child to
helplessness and
learning, the grown man
to hope and communal
purpose: the elder,
loosed from this, goes
back to solitude; yet
you are alone indeed.

Do not think of me as
a prisoner, for I go
abroad, meet men, keep
in touch, and if you
question me you will
find me aware of your
country's happenings no
less than mine, informed
by friends, who will also
finally bury me and take
home my books; the
present is not to die.

347

What sweet cherry-tree
bears leaves, flowers,
hanging fruit, all at
one time?

And between the would-
be hermit, young, and
the actual elder,
the middle span,
the wars?

There my life lay, in
love and the bravery of
power through the world
at large, action out of
the brooding of youth,
only to return to
meditation, with all
those things remembered,
and here begins, late in
most men, a sense of
history, at the inter-
stice of hope and memory.

Memory, so that great
one says, is Prenotion
and Emblem; Thoth in
the Egyptian grot reads
the future in his silky
ibis head.

Strange indeed it is
that now we pursue so
zestfully the extremes
of remoteness, distance,
difficulty, letting that
which is nearest, our
lives and their shape,
the lives of those round
us, of the great human
family, go by default.

Must we not learn to
perceive relations, in
which only a long dynamic
memory, that is knowledge
and insight, avails for
an ordering of events
past and future (yet
neither with over-

literalness nor arbitrary
fantasies) in their
secret sympathy?

And only memory recorded
preserves the past for
the future, or the
treasure of evidence is
lost, tale of each single
life how seeming
insignificant soever,
since none can fail to
glance and reflex that
great movement, sea,
history, the life of the
times.

For these crowding
towards me, multitudes
fresh from the glow of
sleep, lips open on a
thirsty murmur, what
kiss, what speech, in
gift for them?

Given the will to
find meaning in our
own story, there is
need of art by which to
read relationships,
interpretation of signs
and figures, self-
metamorphosis, yet how
little we attain it, a
tardy precept for our
own short span, a scant
formula or two!

Here are the great
seals, primary stamps
in red and blue and
yellow, horned bulls
of Harappa, chariots,
fluttering winged
figures, minted into
money later; you read
them forward.

To relate: a double
sense: connect and tell
over; this knitting

349

self-narration the duty
perhaps of every man if
history is ever to be
embraced and known.

Direct, though, or
indirect, for even as
I speak it occurs to
me: must not a
historian be also
that relation-weaver,
a poet?

And must not the man of
science, he who, too,
relates relations
wavering, admit he
this or no, between
the observed and the
fictive?

We feel after our fate
in the lines of other
lives, and for this
learning it matters
little whether those
figures be real, as we
say, or imagined, so
that they live.

And thus the myth and
metaphor of the poet
with the hypothesis,
hypopoiesis, of the
scientist are truth-
bearing turn by turn;
certainly I have known
the truth of my own
nature unfolded by the
poets' songs.

But look where he draws
the purple cloak over
his face as the bard's
song of heroes and of
a god-willed doom peals
out, lest others mark
the welling from his
deep-salt eyes, though
a lazy glint of gold
like a lovely woman

350

ELIZABETH SEWELL

comes up through the
full red wine in the
bowl beside him and the
mighty feast goes on.

You have been
fortunate enough
to have had poets
by you?

                    Ah, some few, wandering
                    race that they are; yet
                    I too wander, meeting
                    some, to be remembered
                    for ever, in Sweden say,
                    Saxony, or in Illyria.

Continually they set
out, barefoot, into the
mountains, the old man
in white, the boy who
leads him in rough green,
and before them, miles
away, a vast rock-wall
from which the climbers
fall like flies.

And you too have
travelled far and wide,
adding riches as you
went to your own
history.

                    He who seeks to know the
                    ground on which we stand
                    and its as yet inviolate
                    treasure is bound to go
                    abroad through the world.

You and your like are
inverted star-gazers,
it seems: those have
looked to prophesy by
the lights above us,
you to sound under
our careless foot-soles
the illimitable past.

                    And the two meet and
                    fuse in man, as nature
                    in her human ages
                    renounces cataclysm,

351

meteorites no longer
showering nor huge
floods whelming
continents nor the sea-
floor cracked with fire,
and works perhaps as an
artist might towards
the unknowable goal
of all.

Adam and Eve as six-
year-old children lie
naked side by side in
the hot white sand by
a sea of amethyst.

Adventure and action
call us out into the
doings of our time,
and I, far into the
Orient with my love,
thought little of what
underlies man and all
his works as I went
striding mail-clad in
the daylight past the
darker houses of the
rocks, for my love's
arms were my home and
it was there at last I
thought to die; but in
the end to the older darker
house I have come alone,
for the great past of
the mountains hems me
round, and I look to
my own future, a
happy death, and to
the future of man, seen
as if from a great
height, a land of
summer promise.

Out of that plain,
the stiff ranks and
shocks of armed men,
instant sprung; others
have sown also to raise
men and women as bending

352

corn in the valleys
and the water-and-stone
lilies of metropolis.

                                    And now, having
                                    begun, where shall
                                    we end?

Truly there is no
end, only that
peculiar ripeness
which spells, as I
learned in my life,
a new beginning.
God send you all such.

                                    For which aspiration,
                                    as for your courtesy,
                                    we thank you.

VI. THE FIFTH STATION: *The poet encounters one of the books in the cave, and the company make their way back to the world outside*

The poet: The child:

Something has happened: yet all ended smoothly;
Talk done, the two set off for a reconnoitre,
Leaving me to the books: something has happened,
        Something appalling.

Whatever it is, nobody else has noticed.
These are history and poetry books; I turn the pages;
The merchants chat in groups, and the child beside me
        Fingers the zither.

My cheeks feel tight on the bones. This place is icy.
Eyes have the sense that everything is transparent.
How to hide or explain: I think I am going
        Out of my mind?

                    Why do you sit so stiff? That's queer—you shake.
                    Are you, like me, trying to keep awake?

It has to do with the books: yet they seemed so friendly
Under my hand, and they, of course, hand-scripted,
And in text and capital small heavenly-bright
    Illuminations.

What could there be in these? humanity's signets,
Battles depicted, weddings, funerals, shipwrecks,
Palaces, caves, kings, priests, and the humble,
    Hero, and stranger.

Until I was drawn to this one, and something happened.
Latin? Italian? whatever language is it?
I could not read it at all, but found the pictures
    Somehow familiar.

        I think I'll lie on my back now and look at the ceiling.
        Voices grow loud, then soft—it's a funny feeling.

I put my face to the book, to a tiny scene
Showing a group of people; then I got up,
Went closer to the lamp, looked now, and saw
    Myself among them.

Chance resemblance, I thought, but picture on picture
Follows, showing my home, my parents, our travels,
Even this secret cave, with the miner, the hermit:
    There is no mistaking.

Am I dreaming or mad, a life or a story, a method
Common to both? And the book has no title or headings,
And what is the meaning and what this paroxysm
    Of sacred horror . . .

        That wasn't comfortable. Are you restless like me?
        Would you let me lean my head back against your knee?

The book is old, we are dressed in antique garments,
But here I am older—taller, graver perhaps,
At court, on shipboard, laureate—here, oh here
    I hold my beloved.

Adventures thick and fast, wild men in a battle,
A friendly conversation with Moors and Arabs,
A beautiful older man who seems to be
    My friend and mentor.

At the end the pictures are dark, hard to make out,
But oh dearest forms in my dreams, I know and greet you,
So there is joy there too. The end of the book
    Seems to be missing.

    Please put your hand on my head . . . There's a voice that
                                      calls . . .
    Enormous bright-red poppies are growing up the walls . . .

    When Adam was being made, God sat, my mother said,
    A long time with him like this, finishing off his head . . .

The other two men come back—the child is stirring—
I feel ashamed they should know of what has happened.
Steady your face, close up the book, just ask him
    Where does it come from?

'It goes a long way back—how long since I read it!—
Telling the tale, I believe, of a poet's sending,
Of where poetry takes him: and the tongue it speaks
    Is old Provençal.

'What else? It is incomplete—I came upon it
In Jerusalem—it was found among the possessions
Of a friend who died there, and I brought it back
    In memory of him.'

He looks at me piercingly—oh, all is known!
Master your tears. Your fellows are making ready,
Close-banded again; you must bear them this joy which doubles
    Mysterious pain.

Only much later, I think, as we stand collected,
Taking a dear farewell of this man, this vision,
Shall we understand, interpret, put into action
    Our journey's outcome.

VII. ENVOI: *having H.A., W.H.A., J.N., in mind*

Probing the indifferent spaces of our cosmos,
Mechanized, tiny, grimly human figures,
Already ankle-deep in a rotted kingdom,
We suffer, not enact, enormous changes,
Lifting our heads as the recurrent voices
Drift down the world-wind crying revolution;

Perhaps too slackly tuned for revolution,
Nostalgic for a Ptolemaic cosmos
And lullabied by all the lambent voices
Of heaven quiring in concentered figures
Subject only to slow and secular changes
Governed by law in a hierarchic kingdom.

Yet 1642 the first great kingdom
Was toppled by Copernican revolution;
Star-word came down to compass civil changes,
Though history, the metaphor of cosmos,
Still moved immutably in cyclic figures
Where fresh events fulfilled ancestral voices.

Next century brings out the crucial voices:
Federate stars beget a New World kingdom;
Elsewhere, a communal meteor, which disfigures
Henceforth by terror every revolution,
And there is something new under the cosmos,
Explosions now the modulus of changes.

Today what text? another Book of Changes
Far to the east, where long-consistent voices
Take organism as the type of cosmos,
As of a mild and bureaucratic kingdom,
A dancing universe whose revolution
Is mutual order amongst moving figures?

Certain it is we need a shift of figures
To change our thinking as our century changes,
To grasp this paradox of revolution

356

As old and new, active and passive voices,
By innovation to restore the kingdom
And ourselves with it into the living cosmos.

Every cosmos is manned by starry figures.
The password changes—answer those sentinel voices:
We seek a Kingdom. The Word is Revolution.

INDEX

Actus reus, 166, 169
'Aesthetic' science, 26
Aesthetics and value properties, 131–41: art categories, 137–8; 'differentiated ostension', 134; differentiated value events, 132; 'elegance', 133–5, 147; 'form' and 'content', 136–7; perceptual processes and terms, 131–2
Analytic fallacy, 106–7
Analytic philosophy, 141–2: simplism, 141
Aristotle, 11, 220, 230, 291: Scale of Nature, 294
Armstrong, 12
Atomic theory, Dalton's, 104
Atomism, 10, 51, 86–7, 90, 92, 99, 119
Augustine, St., 26
Austin, John, 235
Awareness, 315–16: levels of, 318; subsidiary, 316
Axiological study, 140
Ayer, A. J., 141

Bardeen, John, 19, 89–90, 92
Bayliss, L. E., 114
Behavior, 86: of insects, 110–11, 115; of men and animals, 110, 111, 115; patterns, 112–13
Behaviorism, 18–20, 119–21, 319–20: classical, 120; neo-behaviorism, 120; non-behaviorists, 119; subjective, 120
'Being', 289–91, 296–302, 306–7, 313: and entities, 299–300, 302; common and concrete character, 296–8; concept of of Being in general, 296–8
Being and doing, 165–216: awareness of pain, 186–90; intention, 177–9, 199; *mens rea* and *actus reus*, 166, 169, 185, 200; moral obligation, 207–9; moral significance of dreams, 190–1; ontological foundation of action and responsibility, 194, 197–202; status responsibility, 165, 169–72, 200, 201; voluntary character of moral offenses, 166, 172, 174–8, 180–5, 202, 204–5, 209–10; voluntary responsibility, 165, 170–2, 177, 186, 193, 196–210
Bergson, 205, 298
Berkeley, Bishop, 12
Biochemical specificity, 75–7, 81–5: kinetic, 76–7; of messenger RNA, 81–2; protein synthesis, 79, 81–5; species-specificity, 84–5; static, 75–6
Biology—a molecular science?, 73–101: and atomic theory, 86–7; and superconductivity, 17, 87–92, 98–9; biochemical specificity, 75–7, 81–5; complementarity, 91; DNA, 74–81, 83–5, 91–9; genetic transformation, 84–5; *in vitro* experiments, 79, 81–2, 84, 86, 91, 92; RNA messenger, 81–4, 92; self-duplication of the cell, 74–5, 78, 80, 91
Blondin, 317